高等学校专业英语系列教材

光电信息科学与工程专业英语

Professional English for Optoelectronic Information Science and Engineering

李玲　赵晓辉　张文明　主编
张丽　王虹　参编

化学工业出版社
·北京·

内容简介

《光电信息科学与工程专业英语》是以光电技术行业对复合型人才的培养需求为导向编写的专业英语教材，主要涉及了半导体物理、半导体器件、电磁场和电磁波、激光器和光纤通信等方面的内容。同时，每章还对重点词语和疑难句进行了解析，介绍了科技英语翻译的相关知识，旨在使学生掌握专业基础词汇、提高专业文献的阅读和翻译能力，拓展和深化学生对本专业基础知识和相关技术的了解与掌握。

《光电信息科学与工程专业英语》体系合理、题材广泛、内容丰富，可作为高等院校及科研院所光电信息、电子信息、计算机、应用物理以及科技英语翻译等专业本科生教材，也可作为研究生及留学生教材，还可作为上述领域科研技术人员的参考书。

图书在版编目（CIP）数据

光电信息科学与工程专业英语/李玲，赵晓辉，张文明主编. —北京：化学工业出版社，2020.12（2024.8重印）
高等学校专业英语系列教材
ISBN 978-7-122-38268-9

Ⅰ.①光… Ⅱ.①李…②赵…③张… Ⅲ.①光电子技术-信息技术-英语-高等学校-教材 Ⅳ.①TN2

中国版本图书馆CIP数据核字（2020）第256698号

责任编辑：徐雅妮　任睿婷　　　　装帧设计：李子姮
责任校对：宋　玮

出版发行：化学工业出版社（北京市东城区青年湖南街13号　邮政编码100011）
印　　装：北京七彩京通数码快印有限公司
787mm×1092mm　1/16　印张14　字数344千字　　2024年8月北京第1版第4次印刷

购书咨询：010-64518888　　　　售后服务：010-64518899
网　　址：http://www.cip.com.cn
凡购买本书，如有缺损质量问题，本社销售中心负责调换。

定　　价：49.00元

前言

Preface

21 世纪是国际化的高科技时代和信息化时代，作为国际间交流的重要载体——英语，其作用显得更为重要。专业英语是大学外语教学的重要组成部分，其目的是巩固强化基础英语并进行实践应用，从而掌握科技英语技能，使学生能够熟练阅读国外相关文献，掌握国内外本专业发展的最新动态。本书主要针对光学、光信息科学与技术、光电子技术、光电信息工程等专业开设的专业英语课程而编写，旨在使学生掌握专业基础词汇、提高专业文献的阅读和翻译能力，拓展和深化学生对本专业基础知识和相关技术的了解与掌握。

光电子学是发展十分活跃又尚未完善的一门学科。作者在汇集国内外学者的研究成果和总结作者若干年的教学与科研工作的基础上，编写了本书，力求做到：基本概念清晰，内容深入浅出、易于理解。本书既有基本原理的阐述和必要的理论知识分析与讨论，也有典型应用实例、国内外近期发展状况与趋势。希望在有限的篇幅里让读者对电子与光电子领域的基本原理、主要知识体系及应用有所了解和掌握。同时每章还对重点词语和疑难句进行了解析，以方便读者阅读。

本书内容丰富，选材合理，涵盖光电子学基本理论和实际应用两部分内容。这两方面无所偏重，基本平衡。在基础部分，介绍了半导体物理、半导体器件、全息与傅里叶光学、电磁场与电磁波的基础知识；在应用部分，既介绍了激光器、光纤通信等目前广泛应用的高新技术，又介绍了非线性光学、光子晶体光纤等前沿题目，还涉猎到了光电子应用的核心领域。通过阅读本书可使读者在不知不觉间进入到光电子学研究的前沿。本书由河北大学李玲、赵晓辉、张文明主编，张丽、王虹参与编写。在本书编写过程中，编者还参考了其他教材、文献等资料，在此一并向这些作者表示深深的感谢。

由于本书涉及范围广，加之编者水平有限，书中疏漏和不妥之处在所难免，恳请读者提出宝贵意见，以便完善。

编者

2021 年 1 月

目录

Unit 1
Semiconductor Physics

1.1 Energy Bands and Carrier Concentration

1.1.1 Semiconductor Materials

Solid-state materials can be grouped into three classes—insulators, semiconductors, and conductors. The electrical conductivities σ (and the corresponding resistivities $\rho = 1/\sigma$) are associated with some important materials in each of the three classes. Insulators such as fused quartz and glass have very low conductivity, in the order of 1×10^{-18} to 1×10^{-8} S/cm; and conductors such as aluminum and silver have high conductivities, typically from 10^4 to 10^6 S/cm. Semiconductors have conductivities between those of insulators and those of conductors. The conductivity of a semiconductor is generally sensitive to temperature, illumination, magnetic field, and minute amount of impurity atoms. This sensitivity in conductivity makes the semiconductor one of the most important materials for electronic applications.

The study of semiconductor materials began in the early nineteenth century. Over the years many semiconductors have been investigated. The element semiconductors, those composed of single species of atoms, such as silicon (Si) and germanium (Ge), can be found in Column Ⅳ. However, numerous compound semiconductors are composed of two or more elements. For example, gallium arsenide (GaAs) is a Ⅲ-V compound that is a combination of gallium (Ga) from Column Ⅲ and arsenic (As) from Column V.

Prior to the invention of the bipolar transistor in 1947, semiconductors were used only as two-terminal devices, such as rectifiers and photodiodes. In the early 1950s, germanium was the major semiconductor material. However, germanium was proved unsuitable in many applications because germanium devices exhibited high leakage currents at only moderately elevated temperatures. In addition, germanium oxide is water soluble and unsuited for device fabrication. Since the early 1960s silicon has become a practical substitute and has now virtually supplanted germanium as a material for semiconductor fabrication. The main reasons we now use silicon are that silicon devices exhibit much lower leakage currents, and high-quality silicon dioxide can be grown thermally. There is also an economic consideration. Device grade silicon costs much less than any other semiconductor material. Silicon in the form of silica and silicates comprises 25% of the Earth's crust, and silicon is second only to oxygen in abundance. At present, silicon is one of the most studied elements in the periodic table; and silicon technology is by far the most advanced among all semiconductor technologies.

Many of the compound semiconductors have electrical and optical properties that are absent in silicon. These semiconductors, especially gallium arsenide (GaAs), are used mainly for microwave and photonic applications. Although we do not know as much about the technology of compound semiconductors as we do about that of silicon, compound semiconductor technology has advanced partly because of the advances in silicon technology. In this book we are concerned mainly on device physics and processing technology of silicon and gallium arsenide.

1. 1. 2 Crystal Structure

The semiconductor materials we will study are single crystals, that is, the atoms are arranged in a three-dimensional periodic fashion. The periodic arrangement of atoms in a crystal is called a lattice. In a crystal, an atom never stays far from a single, fixed position. The thermal vibrations associated with the atom are centered about this position. For a given semiconductor, there is a unit cell that is representative of the entire lattice; by repeating the unit cell throughout the crystal, one can generate the entire lattice.

In a simple cubic crystal, each corner of the cubic lattice is occupied by an atom that has six equidistant nearest neighboring atoms. The dimension a is called the lattice constant. Only polonium is crystallized in the simple cubic lattice. A body-centered cubic (bcc) crystal, where in addition to the eight corner atoms, an atom is located at the center of the cube. In a bcc lattice, each atom has eight nearest neighboring atoms. Crystals exhibiting bcc lattices include those of sodium and tungsten. A face-centered cubic (fcc) crystal has one atom at each of the six cubic faces in addition to the eight corner atoms. In an fcc lattice, each atom has 12 nearest neighboring atoms. A large number of elements exhibit the fcc lattice form, including aluminum, copper, gold, and platinum.

The element semiconductors, silicon and germanium, have a diamond lattice structure. This structure also belongs to the cubic-crystal family and can be seen as two interpenetrating fcc sublattices with one sublattice displaced from the other by one quarter of the distance along a diagonal of the cube (i. e., a displacement of $a\sqrt{3}/4$). All atoms are identical in a diamond lattice, and each atom in the diamond lattice is surrounded by four equidistant nearest neighbors that lie at the corners of a tetrahedron. Most of the Ⅲ-V compound semiconductors (e. g., GaAs) have a zinc blende lattice, which is identical to a diamond lattice except that one fcc sublattice has Column Ⅲ atoms (Ga) and the other has Column V atoms (As).

Therefore, the crystal properties along different planes are different, and the electrical and other device characteristics are dependent on the crystal orientation. A convenient method of defining the various planes in a crystal is to use Miller indices. These indices are obtained using the following steps:

① Find the intercepts of the plane on the three Cartesian coordinates in terms of the lattice constant.

② Take the reciprocals of these numbers and reduce them to the smallest three integers having the same ratio.

③ Enclose the result in parentheses (hkl) as the Miller indices for a single plane.

1.1.3 Valence Bonds

As discussed, each atom in a diamond lattice is surrounded by four nearest neighbors. Each atom has four electrons in the outer orbit, and each atom shares these valence electrons with its four neighbors. This sharing of electrons is known as covalent bonding; each electron pair constitutes a covalent bond. Covalent bonding occurs between atoms of the same element, or between atoms of different elements that have similar outer-shell electron configurations. Each electron spends an equal amount of time with each nucleus. However, both electrons spend most of their time between the two nuclei. The force of attraction for the electrons by both nuclei holds the two atoms together. For a zinc blende lattice such as gallium arsenide, the major bonding force is from the covalent bonds. However, gallium arsenide has a slight ionic bonding force that is an electrostatic attractive force between each Ga^- ion and its four neighboring As^+ ions, or between each As^+ ion and its four neighboring Ga^- ions.

At low temperatures, the electrons are bound in their respective tetrahedron lattice; consequently, they are not available for conduction. At high temperatures, thermal vibrations may break the covalent bonds. When a bond is broken, a free electron results that can participate in current conduction. An electron deficiency is left in the covalent bond. This deficiency may be filled by one of the neighboring electrons, which results in a shift of the deficiency location, as from location A to location B. We may therefore consider this deficiency as a particle similar to an electron. This fictitious particle is called a hole. It carries a positive charge and moves, under the influence of an applied electric field, in the direction opposite to that of an electron. The concept of a hole is analogous to that of a bubble in a liquid. Although it is actually the liquid that moves, it is much easier to talk about the motion of the bubble in the opposite direction.

1.1.4 Energy Bands

For an isolated atom, the electrons of the atom can have only discrete energy levels. For example, the energy levels for an isolated hydrogen atom are given by the Bohr model

$$E_h = -m_0 q^4 / (8\varepsilon_0^2 h^2 n^2) = -13.6/n^2 \text{eV}$$

where m_0 is the free-electron mass, q is the electronic charge, ε_0 is the free-space permittivity, h is the Planck constant, and n is a positive integer called the principal quantum number. The discrete energies are −13.6 eV for the ground level ($n=1$), −3.4 eV for the first excited level ($n=2$), etc..

We now consider two identical atoms. When they are far apart, the allowed energy levels for a given principal quantum number (e. g., $n = 1$) consist of one doubly degenerate level, that is, each atom has exactly the same energy (e. g., -13.6 eV for $n = 1$). As the two atoms approach one another, the doubly degenerate energy level will split into two levels by the interaction between the atoms. When we bring N atoms together to form a crystal, the N-fold degenerate energy level will split into N separate but closely spaced levels due to atomic interaction. This results in an essentially continuous band of energy.

The detailed energy band structures of crystalline solids have been calculated using quantum mechanics. Such as a schematic diagram of the formation of a diamond lattice crystal from isolated silicon atoms. Each isolated atom has its discrete energy levels (two levels are shown on the far right of the diagram). As the interatomic spacing decreases, each degenerate energy level splits to form a band. Further decrease in spacing causes the bands originating from different discrete levels to lose their identities and merge together, forming a single band. When the distance between atoms approaches the equilibrium interatomic spacing of the diamond lattice (with a lattice constant of 0.543 nm for silicon), this band splits again into two bands. These bands are separated by a region which designates energies that the electrons in the solid cannot possess. This region is called the forbidden gap, or bandgap E_g. The upper band is called the conduction band, while the lower band is called the valence band.

There are energy band diagrams of three classes of solids—insulators, semiconductors, and conductors. In an insulator such as silicon dioxide (SiO_2), the valence electrons form strong bonds between neighboring atoms. These bonds are difficult to break, and consequently there are no free electrons to participate in current conduction. As shown in the energy band diagram, there is a large bandgap. Note that all energy levels in the valence band are occupied by electrons and all energy levels in the conduction band are empty. Thermal energy or an applied electric field cannot raise the uppermost electron in the valence band to the conduction band. Therefore, silicon dioxide is an insulator, which cannot conduct current.

As discussed, bonds between neighboring atoms in a semiconductor are only moderately strong. Therefore, thermal vibrations will break some bonds. When a bond is broken, a free electron along with a free hole results. The bandgap of a semiconductor is not as large as that of an insulator (e. g., Si with a bandgap of 1.12 eV). Because of this, some electrons will be able to move from the valence band to the conduction band, leaving holes in the valence band. When an electric field is applied, both the electrons in the conduction band and the holes in the valence band will gain kinetic energy and conduct electricity.

In conductors such as metals, the conduction band either is partially filled or overlaps the valence band so that there is no bandgap. As a consequence, the uppermost electrons in the partially filled band or electrons at the top of the valence band can move to the next-higher available energy level when they gain kinetic energy (e. g., from an applied electric field). Therefore, current conduction can readily occur in conductors.

The energy band diagrams indicate electron energies. When the energy of an electron is increased, the electron moves to a higher position in the band diagram. On the other hand, when the energy of a hole is increased, the hole moves downward in the valence band. (This is because a hole has a charge opposite that of an electron.)

As we discussed before, the separation between the energy of the lowest conduction band and that of the highest valence band is called the bandgap E_g, which is the most important parameter in semiconductor physics. We designate E_c as the bottom of the conduction band; E_c corresponds to the potential energy of an electron, that is, the energy of a conduction electron that is at rest. The kinetic energy of an electron is measured upward from E_c. Similarly, we designate E_v, as the top of the valence band; E_v corresponds to the potential energy of a hole. The kinetic energy of a hole is measured downward from E_v.

At room temperature and under normal atmosphere, the values of the bandgap are 1.12 eV for silicon and 1.42 eV for gallium arsenide. The bandgap approaches 1.17 eV for silicon and 1.52 eV for gallium arsenide at 0 K.

1.1.5 Density of States

When electrons move back and forth along the x-direction in a semiconductor material, the movements can be described by standing-wave oscillations. The wavelength λ of a standing wave is related to the length of the semiconductor L by

$$\frac{L}{\lambda} = n_x \tag{1.1}$$

where n_x, is an integer. The wavelength can be expressed as

$$\lambda = \frac{h}{\bar{p}_x} \tag{1.2}$$

where h is the Planck constant and $\bar{p}_x$ is the crystal momentum in the x-direction. Substituting Eq. (1.2) into Eq. (1.1) gives

$$L\bar{p}_x = hn_x \tag{1.3}$$

The incremental momentum $\mathrm{d}\bar{p}$ required for a unity increase in n_x is

$$L\mathrm{d}\bar{p}_x = h \tag{1.4}$$

For a three-dimensional cube of side L, we have

$$L^3\mathrm{d}\bar{p}_x\mathrm{d}\bar{p}_y\mathrm{d}\bar{p}_z = h^3 \tag{1.5}$$

The volume $\mathrm{d}\bar{p}_x\mathrm{d}\bar{p}_y\mathrm{d}\bar{p}_z$ in the momentum space for a unit cube ($L = 1$) is thus equal to h^3. Each incremental change in n corresponds to a unique set of integers (n_x, n_y, n_z), which in turn corresponds to an allowed energy state. Thus, the volume in momentum space for an energy state is h^3. The volume between two concentric spheres (from $\bar{p}$ to $\bar{p}+\mathrm{d}\bar{p}$) is $4\pi\bar{p}^2\mathrm{d}\bar{p}$. The number of energy states contained in this volume is then $2(4\pi\bar{p}^2\mathrm{d}\bar{p})/h^3$, where the factor 2 accounts for the electron spins. We can substitute E for $\bar{p}$ and obtain

$$N(E)\mathrm{d}E = \frac{8\pi\bar{p}^2\mathrm{d}\bar{p}}{h^3} = 4\pi\left(\frac{2m_n}{h^2}\right)^{3/2} E^{1/2}\mathrm{d}E \tag{1.6}$$

and

$$N(E)=4\pi\left(\frac{2m_{\mathrm{n}}}{h^{2}}\right)^{3/2}E^{1/2} \tag{1.7}$$

where $N(E)$ is called the density of states, that is, the density of allowed energy states per unit volume.

1.1.6 Intrinsic Carrier Concentration

At finite temperatures, continuous thermal agitation results in the excitation of electrons from the valence band to the conduction band and leaves an equal number of holes in the valence band. An intrinsic semiconductor is one that contains relatively small amounts of impurities compared to the thermally generated electrons and holes. To obtain the electron density (i. e., the number of electrons per unit volume) in an intrinsic semiconductor, we first evaluate the electron density in an incremental energy range dE. This density $n(E)$ is given by the product of the density of allowed energy states per unit volume $N(E)$ and by the probability of occupying that energy range $F(E)$. Thus, the electron density in the conduction band is given by integrating $N(E)F(E)\mathrm{d}E$ from the bottom of the conduction band (E_{c} initially taken to be $E=0$ for simplicity) to the top of the conduction band E_{top}

$$n=\int_{0}^{E_{\mathrm{top}}}n(E)\mathrm{d}E=\int_{0}^{E_{\mathrm{top}}}N(E)F(E)\mathrm{d}E \tag{1.8}$$

The probability that an electronic state with energy E is occupied by an electron given by the Fermi-Dirac distribution function (which is also referred to as the Fermi distribution function)

$$F(E)=\frac{1}{1+\exp(E-E_{\mathrm{F}})/kT} \tag{1.9}$$

where k is the Boltzmann constant, T is the absolute temperature in degrees Kelvin, and E_{F} is the Fermi level. The Fermi level is the energy at which the probability of occupation by an electron is exactly one half. The Fermi distribution function can thus be approximated by simpler expressions

$$F(E)\approx\exp(E-E_{\mathrm{F}})/kT$$

for

$$(E-E_{\mathrm{F}})>3kT \tag{1.10a}$$

and

$$F(E)\approx\exp[-(E_{\mathrm{F}}-E)/kT]$$

for

$$(E-E_{\mathrm{F}})<-3kT \tag{1.10b}$$

Eq. (1.10b) can be regarded as the probability that a hole occupied a state located at energy E.

In the conduction band there are a large number of allowed states. However, for intrinsic semiconductors there will be only a few electrons in the conduction band, hence the probability of an electron occupying one of these states is small. There are also a large number of allowed states in the valence band. By contrast most of those are occupied by electrons. Thus, the probability of an electron occupying one of these states in the valence band is nearly unity. There will be only a few unoccupied electron states, that is, holes, in the valence band. As can be seen, the Fermi level is located near the middle of the bandgap (i. e., E_F is many kT below E_c). Because $F(E)$ is an exponentially decreasing function of E[Eq. (1. 10a)], the value of E_{top} can be replaced by infinity. Substituting Eq. (1. 10a) into Eq. (1. 8) yields

$$n = 4\pi\left(\frac{2m_n}{h^2}\right)^{3/2}\int_0^{\infty} E^{1/2}\exp\left(-\frac{E-E_F}{kT}\right)dE \tag{1. 11}$$

and

$$N_c \equiv 2\left(\frac{2\pi m_n kT}{h^2}\right)^{3/2}$$

We obtain the electron density in the conduction band where N_c is the effective density of states in the conduction band. At room temperature (300 K), N_c is 2.8×10^{19} cm^{-3} for silicon and $4.7\times10^{17} cm^{-3}$ for gallium arsenide.

Similarly, we can obtain the hole density p in the valence band.

$$N_v = 2\left(\frac{2\pi m_p kT}{h^2}\right)^{3/2}$$

where N_v, is the effective density of states in the valence band. At room temperature, N_v is 1.04×10^{19} cm^{-3} for silicon and 7.0×10^{18} cm^{-3} for gallium arsenide.

For an intrinsic semiconductor as defined previously, the number of electrons per unit volume in the conduction band is equal to the number of holes per unit volume in the valence band, that is, $n=p=n_i$, where n_i is the intrinsic carrier density. Note that the shaded area in the conduction band is the same as that in the valence band.

At room temperature, the second term is much smaller than the bandgap. Hence, the intrinsic Fermi level E_i, that is, the Fermi level of an intrinsic semiconductor, generally lies very close to the middle of the bandgap.

The intrinsic carrier density is obtained

$$n_i = \sqrt{N_c N_v}\exp\left(-\frac{E_g}{kT}\right)$$

$$np = n_i^2 \tag{1. 12}$$

where $E_g = E_c - E_v$. Eq. (1. 12) is called the mass action law, which is valid for both intrinsic and extrinsic (i. e., doped with impurities) semiconductors under a thermal equilibrium condition. In an extrinsic semiconductor, the increase of one type of carrier tends to reduce the number of the other type through recombination; thus, the product of the two types of carriers will remain constant at a given temperature.

1. 1. 7 Donors and Acceptors

When a semiconductor is doped with impurities, the semiconductor becomes extrinsic and impurity energy levels are introduced. A silicon atom is replaced (or substituted) by an arsenic atom with five valence electrons. The arsenic atom forms covalent bonds with its four neighboring silicon atoms. The fifth electron becomes a conduction electron that is "donated" to the conduction band. The silicon becomes n-type because of the addition of the negative charge carrier, and the arsenic atom is called a donor. Similar, when a boron atom with three valence electrons substitute for a silicon atom, additional electron is accepted to form four covalent bonds around the boron, and a positively charged "hole" is created in the valence band. This is a p-type semiconductor, and the boron is an acceptor.

For shallow donors in silicon and gallium arsenide, there usually is enough thermal energy to supply the energy E_D to ionize all donor impurities at room temperature and thus provide an equal number of electrons in the conduction band. This condition is called complete ionization. Under a complete ionization condition, we can write the electron density as

$$n = N_D$$

where N_D is the donor concentration. We obtain the Fermi level in term of the effective density of states N_c and the donor concentration N_D

$$E_c - E_F = kT\ln(N_c / N_D)$$

From equation it is clear that the higher the donor concentration, the smaller the energy difference $(E_c - E_F)$, that is, the Fermi level will move closer to the bottom of the conduction band. Similarly, for higher acceptor concentrations, the Fermi level will move closer to the top of the valence band. However, the Fermi level is closer to the bottom of the conduction band, and the electron concentration (upper shaded area) is much larger than the hole concentration (lower shaded area). Because of the mass action law, the product of n and p is the same as that for the intrinsic case (i. e., $np = n_i^2$).

1.2 Carrier Transport Phenomena

In this section we consider various transport phenomena that arise from the motion of charge carriers (electrons and holes) in semiconductors under the influence of an electric field and a carrier concentration gradient. We shall discuss the concept of injection of excess carriers, which gives rise to a non-equilibrium condition, that is, the carrier concentration product pn is different from its equilibrium value n_i^2. Return to an equilibrium condition through the generation-recombination processes will be considered next. We shall then derive the basic governing equations for semiconductor device operation, which include the current density equation and the continuity equation. The section closes with a brief discussion of high-field effects, which result in velocity saturation and impact ionization.

1.2.1 Carrier Drift

Consider an n-type semiconductor sample with uniform donor concentration in thermal equilibrium. As discussed, the conduction electrons in the semiconductor conduction band are essentially free particles, since they are not associated with any particular lattice or donor site. The influence of crystal lattices is incorporated, in the effective mass of conduction electrons, which differs somewhat from the mass of free electrons. Under thermal equilibrium, the average thermal energy of a conduction electron can be obtained from the theorem for equipartition of energy: 1/2 kT units of energy per degree of freedom, where k is Boltzmann constant and T is the absolute temperature. The electrons in a semiconductor have three degrees of freedom; they can move about in a three-dimensional space. Therefore, the kinetic energy of the electrons is given by

$$\frac{1}{2}m_n v_{th}^2 = \frac{3}{2}kT \quad (1.13)$$

where m_n is the effective mass of electrons and v_{th} is the average thermal velocity. At room temperature (300 K), the thermal velocity is about 10^7 cm/s for silicon and gallium arsenide.

The electrons in the semiconductor are therefore moving rapidly in all directions. The thermal motion of an individual electron may be visualized as a succession of random scattering from collisions with lattice atoms, impurity atoms, and other scattering centers. The random motion of electrons leads to a zero net displacement of an electron over a sufficiently long period of time. The average distance between collisions is called the mean free path, and the average time between collisions is called the mean free time τ_c. For a typical value of 10^{-5} cm for the mean free path, τ_c is about 1 ps (i. e., $10^{-5}/v_{th} \approx 10^{-12}$ s).

When a small electric field E is applied to the semiconductor sample, each electron will experience a force $-qE$ from the field and will be accelerated along the field (in the opposite direction to the field) during the time between collisions. Therefore, an additional velocity component will be superimposed upon the thermal motion of electrons. This additional component is called the drift velocity. The combined displacement of an electron is due to the random thermal motion. Note that there is a net displacement of the electron in the direction opposite to the applied field.

We can obtain the drift velocity v, by equating the momentum (force × time) applied to an electron during the free flight between collisions to the momentum gained by the electron in the same period. The equality is valid because in a steady state all momentum gained between collisions is lost to the lattice in the collision. The momentum applied to an electron is given by $-qE$, and the momentum gained is $m_n v_n$, we have

$$-q\varepsilon\tau_c = m_n v_n \quad (1.14)$$

or

$$v_n = -\left(\frac{q\tau_c}{m_n}\right)\varepsilon \quad (1.14a)$$

Eq. (1.14a) states that the electron drift velocity is proportional to the applied electric field. The proportionality factor depends on the mean free time and the effective mass. The proportionality factor is called the electron mobility μ_n, in units of $cm^2/(V \cdot s)$, or

$$\mu_n \equiv \frac{q\tau_c}{m_n} \quad (1.15)$$

Thus

$$v_n = -\mu_n \varepsilon \tag{1.16}$$

Mobility is an important parameter for carrier transport because it describes how strongly the motion of an electron is influenced by an applied electric field. A similar expression can be written for holes in the valence band

$$v_p = \mu_p \varepsilon \tag{1.17}$$

where v_p is the hole drift velocity and μ_p is the hole mobility. The negative sign is removed in Eq. (1.17) because holes drift in the same direction as the electric field.

In Eq. (1.15) the mobility is related directly to the mean free time between collisions, which in turn is determined by the various scattering mechanisms. The two most important mechanisms are lattice scattering and impurity scattering. Lattice scattering results from thermal vibrations of the lattice atoms at any temperature above absolute zero. These vibrations disturb lattice periodic potential and allow energy to be transferred between the carriers and the lattice. Since lattice vibration increases with increasing temperature, lattice scattering becomes dominant at high temperatures; hence the mobility decreases with increasing temperature. Theoretical analysis shows that the mobility due to lattice scattering μ_L will decrease in proportion to $T^{-3/2}$.

Impurity scattering results when a charge carrier travels past an ionized dopant impurity (donor or acceptor). The charge carrier path will be deflected due to Coulomb force interaction. The probability of impurity scattering depends on the total concentration of ionized impurities, that is, the sum of the concentration of negatively and positively charged ions. However, unlike lattice scattering, impurity scattering becomes less significant at higher temperatures. At higher temperatures, the carriers move faster; they remain near the impurity atom for a shorter time and are therefore less effectively scattered. The mobility due to impurity scattering μ_I can theoretically be shown to vary as $T^{3/2}/N_T$, where N_T is the total impurity concentration.

The probability of a collision taking place in unit time, $1/\tau_c$, is the sum of the probabilities of collisions due to the various scattering mechanisms.

The measured electron mobility as a function of temperature for silicon with five different donor concentrations is shown. The insert shows the theoretical temperature dependence of mobility due to both lattice and impurity scatterings. For lightly doped samples (e.g., the sample with doping of 10^{14} cm^{-3}), the lattice scattering dominates, and the mobility decreases as the temperature increases. For heavily doped samples, the effect of impurity scattering is most pronounced at low temperatures. The mobility increases as the temperature increases, as can be seen for the sample with doping of 10^{19} cm^{-3}. For a given temperature, the mobility decreases with increasing impurity concentration, because of enhanced impurity scatterings.

The measured mobilities in silicon and gallium arsenide as a function of impurity concentration at room temperature are shown. Mobility reaches a maximum value at low impurity concentrations; this corresponds to the lattice-scattering limitation. Both electron and hole mobilities decrease with increasing impurity concentration and eventually approach a minimum value at high concentrations. Note also that the mobility of electrons is greater than that of holes. Greater electron mobility is due mainly to the smaller effective mass of electrons.

1.2.2 Carrier Diffusion

In the preceding section we considered the drift current, that is, the transport of carriers when an electric field is applied. Another important current component can exist if there is a spatial variation of carrier concentration in the semiconductor material, that is, the carriers tend to move from a region of high concentration to a region of low concentration. This current component is called diffusion current.

To understand the diffusion process, let us assume an electron density that varies in the x-direction. The semiconductor is at uniform temperature, so that the average thermal energy of electrons does not vary with x, only the density $n(x)$ varies. We shall consider the number of electrons crossing the plane at $x=0$ per unit time and per unit area. Because of finite temperature, the electrons have random thermal motions with a thermal velocity v_{th} and a mean free path l. (Note that $l=v_{\text{th}}\tau_{\text{c}}$, where τ_{c} is the mean free time.) The electrons at $x=-l$, one mean free path away on the left side, have equal chances of moving left or right; and in a mean free time τ_{c}, one half of time will move across the plane $x=0$.

Because each electron carries a charge $-q$, the carrier flow gives rise to a current

$$J_{\text{n}}=qD_{\text{n}}\frac{\mathrm{d}n}{\mathrm{d}x} \tag{1.18}$$

where $D_{\text{n}}\equiv v_{\text{th}}l$ is called diffusivity.

The diffusion current is proportional to the spatial derivative of the electron density. Diffusion current results from the random thermal motion of carriers in a concentration gradient. For an electron density that increases with x, the gradient is positive, and the electrons will diffuse toward the negative x-direction. The current is positive and flows in the direction opposite to that of the electrons.

1.2.3 Carrier Injection

In thermal equilibrium the relationship $pn=n_{\text{i}}^2$ is valid. If excess carriers are introduced to a semiconductor so that $pn>n_{\text{i}}^2$, we have a none-equilibrium situation. The process of introducing excess carriers is called carrier injection. We can inject carriers by using various methods including optical excitation and forward biasing a p-n junction. In the case of optical excitation, we shine a light on a semiconductor. If the photon energy $h\nu$ of the light is greater than the bandgap energy E_{g} of the semiconductor, where h is the Planck constant and ν is the optical frequency, the photon is absorbed by semiconductor and an electron-hole pair is generated. The optical excitation increases the electron and hole concentrations above their equilibrium values. These additional carriers are called excess carriers.

The magnitude of the excess carrier concentration relative to the majority carrier concentration determines the injection level. We shall use an example to clarify the meaning of injection level. The majority carrier concentration is approximately equal to the donor concentration, that is, $n_{n0} = 10^{15}\ cm^{-3}$. The minority carrier concentration is given by $p_{n0} = n_i^2 / n_{n0}$, $= 1.45 \times 10^5\ cm^{-3}$. In this notation, the first subscript refers to the type of semiconductor and the subscript 0 refers to the thermal equilibrium condition. Thus, n_{n0} and p_{n0} denote the electron and hole concentrations, respectively, in an n-type semiconductor in equilibrium.

When we introduce (e. g., by optical excitation) excess carriers of both types into the semiconductor, the excess electron concentration Δn must equal the excess hole concentration Δp, because electrons and holes are produced in pairs. For example, we have added $10^{12}\ cm^{-3}$ minority carriers (holes in an n-type semiconductor). Therefore, the hole concentration is increased by seven orders of magnitude (from $10^5\ cm^{-3}$ to $10^{12}\ cm^{-3}$). At the same time we have added $10^{12}\ cm^{-3}$ majority carriers (electrons) to the semiconductor. However, this concentration of excess electrons is negligibly small compared to the original electron concentration. The percentage change in the majority carrier concentration is only 0. 1%. This condition, in which the excess carrier concentration is small in comparison to the doping concentration, that is, $\Delta n = \Delta p \ll N_D$, is referred to as low-level injection.

As an example, high-level injection in which the number of injected excess carriers is comparable to or larger than the number of carriers due to doping concentration. In this case, the injected-carrier concentrations may overwhelm the equilibrium majority carrier concentration, and p, becomes comparable to n, as indicated in the figure. High-level injection is sometime encountered in device operations. However, because of the complexities involved in its treatment, we shall be concerned mainly with low-level injection.

1. 2. 4 Generation and Recombination Processes

Whenever the thermal-equilibrium condition is disturbed (i. e., $pn \neq n_i^2$), processes exist to restore the system to equilibrium (i. e., $pn = n_i^2$). In the case of the injection of excess carriers, the mechanism that restores equilibrium is recombination of the injected minority carriers with the majority carriers. Depending on the nature of the recombination process, the released energy that results from the recombination process can be emitted as a photon or dissipated as heat to the lattice. When a photon is emitted, the process is called radiative recombination; otherwise, it is called nonradiative recombination.

Recombination phenomena can be classified as direct and indirect processes. Direct recombination, also called band-to-band recombination, usually dominates in direct-bandgap semiconductors, such as gallium arsenide; while indirect recombination via bandgap recombination centers dominates in indirect-bandgap semiconductors, such as silicon.

Consider a direct-bandgap semiconductor in thermal equilibrium. The continuous thermal vibration of lattice atoms causes some bonds between neighboring atoms to be broken. When a bond is broken, an electron-hole pair is generated. In terms of the band diagram, the thermal energy enables a valence electron to make an upward transition to the conduction band leaving a hole in the valence band. This process is called carrier generation and is represented by the generation rate G_{th} (number of electron-hole pairs generated per cm^3 per second). When an electron makes a transition downward from the conduction band to the valence band, an electron-hole pair is annihilated. This reverse process is called recombination; it is represented by the recombination rate R_{th}. Under thermal equilibrium conditions, the generation rate G_{th} must equal the recombination rate R_{th}, so that the carrier concentrations remain constant and the condition $pn = n_i^2$ is maintained.

When excess carriers are introduced to a direct-bandgap semiconductor, the probability is high that electrons and holes will recombine directly, because the bottom of the conduction band and the top of the valence band are lined up and no additional crystal momentum is required for the transition across the bandgap. The rate of the direct recombination R is expected to be proportional to the number of holes available in the valence band; that is

$$R = \beta np$$

where β is the proportionality constant. As discussed previously, in thermal equilibrium the recombination rate must be balanced by the generation rate. Therefore, for an n-type semiconductor, we have

$$G_{th} = R_{th} = \beta n_{n0} p_{n0}$$

where n_{n0} and p_{n0} represent electron and hole densities in an n-type semiconductor at thermal equilibrium. When we shine a light on the semiconductor to produce electron-hole pairs at a rate G_L, the carrier concentrations are above their equilibrium values.

Therefore, the net recombination rate is proportional to the excess minority carrier concentration. Obviously, $U = 0$ in thermal equilibrium. The proportionality constant $1/\beta n_{n0}$ is called the lifetime τ_p of the excess minority carriers. The physical meaning of lifetime can best be illustrated by the transient response of a device after the sudden removal of the light source. Consider an n-type sample that is illuminated with light and in which the electron-hole pairs are generated uniformly throughout the sample with a generation rate G_L. As for the variation of p_n with time. The minority carriers recombine with majority carriers and decay exponentially with a time constant τ_p which corresponds to the lifetime.

The above case illustrates the main idea of measuring the carrier lifetime using photoconductivity method. The excess carriers, generated uniformly throughout the sample by the light pulse, cause a momentary increase in the conductivity. The increase in conductivity manifests itself by a drop in voltage across the sample when a constant current passes through it. The decay of the conductivity can be observed on an oscilloscope and is a measure of the lifetime of the excess minority carriers.

1.3 PN Junction

Most semiconductor devices contain at least one junction between p-type and n-type material. These p-n junctions are fundamental to the performance of functions such as rectification, amplification, switching, and other operations in electronic circuits. In this section we shall discuss the equilibrium state of junction and the flow of electrons and holes across a junction under steady state and transient conditions.

1.3.1 Equilibrium Conditions

In this section we wish to develop both a useful mathematical description of the p-n junction and a strong qualitative understanding of its properties. There must be some compromise in these two goals, since a complete mathematical treatment would obscure the essentially simple physical features of junction operation; on the other hand, a completely qualitative description would not be useful in making calculations. The approach, therefore, will be to describe the junction mathematically while neglecting small effects, which add little to the basic solution.

The mathematics of p-n junctions is greatly simplified for the case of the step junction, which has uniform p doping on one side of a sharp junction and uniform n doping on the other side. This model represents the alloyed junction quite well; the diffused junction, however, is actually graded ($N_d - N_a$ varies over a significant distance on either side of the junction). After the basic ideas of junction theory are explored for the step junction, we can make the appropriate corrections to extend the theory to the graded junction. In these discussions we shall assume one-dimensional current flow in samples of uniform cross-sectional area.

In this section we investigate the properties of the step junction at equilibrium (i. e., with no external excitation and no net currents flowing in the device). We shall find that the difference in doping on each side of the junction causes a potential difference between the two types of material. This is a reasonable result, since we would expect some charge transfer because of diffusion between the p material (many holes) and the n material (many electrons). In addition, we shall find that there are four components of current, which flow across the junction due to the drift and diffusion of electrons and holes. These four components combine to give zero net current for the equilibrium case. However, the application of bias to the junction increases some of these current components with respect to others, giving net current flow. If we understand the nature of these four current components, a sound view of p-n junction operation, with or without bias, will follow.

Let us consider separate regions of p and n-type semiconductor material, brought together to form a junction. This is not a practical way of forming a device, but it does allow us to discover the requirements of equilibrium at a junction. Before they are joined, the n material has a large concentration of electrons and few holes, whereas the converse is true for the p material. Upon joining the two regions, we expect diffusion of carriers to take place because of the large carrier concentration gradients at the junction. Thus holes diffuse from the p side into the n side, and electrons diffuse from n to p. The resulting diffusion current cannot build up indefinitely, however, because an opposing electric field is created at the junction. If the two regions were boxes of red air molecules and green molecules (perhaps due to appropriate types of pollution), eventually there would be a homogeneous mixture of the two after the boxes were joined. This cannot occur in the case of the charged particles in a p-n junction because of the development of space charge and the electric field. If we consider that electrons diffusing from n to p leave behind uncompensated donor ions (Nd^+) in the n material, and holes leaving the p region leave behind uncompensated acceptors (Na^-), it is easy to visualize the development of a region of positive space charge near the n side of the junction and negative charge near the p side. The resulting electric field is directed from the positive charge toward the negative charge. Thus ε is in the direction opposite to that of diffusion current for each type of carrier (recall electron current is opposite to the direction of electron flow). Therefore, the field creates a drift component of current from n to p, opposing the diffusion current.

Since we know that no net current can flow across the junction at equilibrium, the current due to the drift of carriers in the E field must exactly cancel the diffusion current. Furthermore, since there can be no net buildup of electrons or holes on either side as a function of time, the drift and diffusion currents must cancel for each type of carrier

$$J_p(\text{drift}) + J_p(\text{diff.}) = 0$$

$$J_n(\text{drift}) + J_n(\text{diff.}) = 0$$

Therefore, the electric field builds up to the point where the net current is zero at equilibrium. The electric field appears in some region W about the junction, and there is an equilibrium potential difference V_0 across W. In the electrostatic potential diagram, there is a gradient in potential in the direction opposite to ε, in accordance with the fundamental relation. $\varepsilon(x) = \mathrm{d}v(x)/\mathrm{d}x$. We assume the electric field is zero in the neutral regions outside W. Thus there is a constant potential v_n in the neutral n material, a constant v_p in the neutral p material, and a potential difference $V_0 = v_n - v_p$ between the two. The region W is called the transition region, and the potential difference V_0 is called the contact potential. The contact potential appearing across W is a built-in potential barrier, in that it is necessary to the maintenance of equilibrium at the junction; it does not imply any external potential. Indeed, the contact potential cannot be measured by placing a voltmeter across the device, because new contact potentials are formed at each probe, just canceling V_0. By definition V_0 is an equilibrium quantity, and no net current can result from it.

The contact potential separates the bands; the valence and conduction energy bands are higher on the p side of the junction than on the n side by the amount qV_0. The separation of the bands at equilibrium is just that required to make the Fermi level constant throughout the device. We shall show that this is the case in the following section. However, we could predict the lack of spatial variation of the Fermi level from thermodynamic arguments. Any gradient in the quasi-Femi level implies a net current. Since $E_F = F_n = F_p$ at equilibrium, and since the net current must be zero, we conclude that E_F must be constant across the junction. Thus if we know the band separation for the junction simply by drawing a diagram with the Fermi levels aligned.

To obtain a quantitative relationship between V_0 and the doping concentrations on each side of the junction, we must use the requirements for equilibrium in the drift and diffusion current equations. The drift and diffusion components of the hole current just cancel at equilibrium

$$J_p(x) = q\left[\mu_p p(x)\varepsilon(x) - D_p \frac{dp(x)}{dx}\right] = 0 \tag{1.19}$$

1.3.2 Forward And Reverse Biased Junctions

One useful feature of a p-n junction is that current flows quite freely in the p to n direction when the p region has a positive external voltage bias relative to n (forward bias and forward current), whereas virtually no current flows when p is made negative relative to n (reverse bias and reverse current). This asymmetry of the current flow makes the p-n junction diode very useful as a rectifier. As an example of rectification, suppose a sinusoidal a-c generator is placed in series with a resistor and a diode, which passes current only in one direction. The resulting current through the resistor will reflect only half of the sinusoidal signal, for example, only the positive half-cycles. The rectified sine wave has an average value and can be used, for example, to charge a storage battery; on the other hand, the input sine wave has no average value. Diode rectifiers are useful in many applications in electronic circuits, particularly in "wave-shaping" (making use of the nonlinear nature of the diode to change the shape of time-varying signals).

While rectification is an important application, it is only the beginning of a host of uses for the biased junction. Biased p-n junctions can be used as voltage-variable capacitors, photocells, light emitters, and many more devices, which are basic to modern electronics. Two or more junctions can be used to form transistors and controlled switches.

In this section we begin with a qualitative description of current flow in a biased junction. With the background of the previous section, the basic features of current flow are relatively simple to understand, and these qualitative concepts form the basis for the analytical description of forward and reverse currents in a junction.

We assume that an applied voltage bias V appears across the transition region of the junction rather than in the neutral n and p regions. Of course, there will be some voltage drop in the neutral material, if a current flows through it. But in most p-n junction devices, the length of each region is small compared with its area, and the doping is usually moderate to heavy; thus the resistance is small in each neutral region, and only a small voltage drop can be maintained outside the space charge (transition) region. For almost all calculations it is valid to assume that an applied voltage appears entirely across the transition region. We shall take V to be positive when the external bias is positive on the p side relative to the n side.

Since an applied voltage changes the electrostatic potential barrier and thus the electric field within the transition region, we would expect changes in the various components of current at the junction. In addition, the separation of the energy bands is affected by the applied bias, along with the width of the depletion region. Let us begin by examining qualitatively the effects of bias on the important features of the junction.

The electrostatic potential barrier at the junction is lowered by a forward bias V_f from the equilibrium contact potential V_0 to the smaller value $V_0 - V_f$. This lowering of the potential barrier occurs because a forward bias (p positive with respect to n) raises the electrostatic potential on the p side relative to the n side. For a reverse bias ($V = -V_r$) the opposite occurs; the electrostatic potential of the p side is depressed relative to the n side, and the potential barrier at the junction becomes larger ($V_0 + V_r$).

The electric field within the transition region can be deduced from the potential barrier. We notice that the field decreases with forward bias, since the applied electric field opposes the built-in field. With reverse bias the field at the junction is increased by the applied field, which is in the same direction as the equilibrium field.

The change in electric field at the junction calls for a change in the transition width W, since it is still necessary that a proper number of positive and negative charges (in the form of uncompensated donor and acceptor ions) are exposed for a given value of the field. Thus we would expect the width W to decrease under forward bias (smaller ε, fewer uncompensated charges) and to increase under reverse bias.

The separation of the energy bands is a direct function of the electrostatic potential barrier at the junction. The height of the electron energy barrier is simply the electronic charge q times the height of the electrostatic potential barrier. Thus the bands are separated less $q(V_0 - V_f)$ under forward bias than at equilibrium, and more under reverse bias $q(V_0 - V_r)$. We assume the Fermi level deep inside each neutral region is essentially the equilibrium value (we shall return to this assumption later); therefore, the shifting of the energy bands under bias implies a separation of the Fermi levels on either side of the junction. Under forward bias, the Fermi level on the n side E_{Fn} is above E_{Fp} by the energy qV_f; for reverse bias, E_{Fp} is qV_r, joules higher than E_{Fn}. In energy units of electron volts, the Fermi levels in the two neutral regions are separated by applied voltage.

The diffusion current is composed of majority carrier electrons on the n side surmounting the potential energy barrier to diffuse to the p side, and holes surmounting their barrier from p to n. There is a distribution of energies for electrons in the n-side conduction band, and some electrons in the high-energy "tail" of the distribution have enough energy to diffuse from n to p at equilibrium in spite of the barrier. With forward bias, however, the barrier is lowered to ($V_0 - V_f$), and many more electrons in the n-side conduction band have sufficient energy to diffuse from n to p quite large with forward bias. Similarly, more holes can diffuse from p to n under forward bias because of the lowered barrier. For reverse bias the barrier becomes so large ($V_0 + V_f$) that virtually no electrons in the n-side conduction band or holes in the p-side valence band have enough energy to surmount it. Therefore, the diffusion current is usually negligible for reverse bias.

The drift current is relatively insensitive to the height of the potential barrier. This sounds strange at first, since we normally think in terms of material with ample carriers, and therefore we expect drift current to be simply proportional to the applied field. The reason for this apparent anomaly is the fact that the drift current is limited not by how fast carriers are swept down the barrier, but rather how often. For example, minority carrier electrons on the p side, which wander into the transition region will be swept down the barrier by the field, giving rise to the electron component of drift current. However, this current is small not because of the size of the barrier, but because there are very few minority electrons in the p side to participate. Every electron on the p side, which diffuses to the transition region will be swept down the potential energy hill, whether the hill is large or small. The electron drift current does not depend on how fast an individual electron is swept from p to n, but rather on how many electrons are swept down the barrier per second. Similar comments apply regarding the drift of minority holes from the n side to the p side of the junction. To a good approximation, therefore, the electron and hole drift currents at the junction are independent of the applied voltage.

The supply of minority carriers on each side of the junction required to participate in the drift component of current is generated by thermal excitation of electron-hole pairs. For example, an EHP created near the junction on the p side provides a minority electron in the p material. If the EHP is generated within a diffusion length L_n of the transition region, this electron can diffuse to the junction and be swept down the barrier to the n side. The resulting current due to drift of generated carriers across the junction is commonly called the generation current since its magnitude depends entirely on the rate of generation of EHPs. As we shall discuss later, this generation current can be increased greatly by optical excitation of EHPs near the junction (the p-n junction photodiode).

The total current crossing the junction is composed of the sum of the diffusion and drift components: the electron and hole diffusion currents are both directed from p to n (although the particle flow directions are opposite to each other), and the drift currents are from n to p. The net current crossing the junction is zero at equilibrium, since the drift and diffusion components cancel for each type of carrier (the equilibrium electron and hole components do not need to be equal, as long as the net hole current and the net electron current are both zero). Under reverse bias, both diffusion components are negligible because of the large barrier at the junction, and the only current is the relatively small (and essentially voltage-independent) generation current from n to p. This generation current is in a sketch of a typical I-V plot for a p-n junction. In this figure, the positive direction for the current I is taken from p to n, and the applied voltage V is positive when the positive battery terminal is connected to p and the negative terminal to n. The only current flowing in this p-n junction diode for negative V is the small current I(gen.) due to carriers generated in the transition region or minority carriers which diffuse to the junction and are collected. The current at $V=0$ (equilibrium) is zero since the generation and diffusion currents cancel

$$I = I(\text{diff.}) - |I(\text{gen.})| = 0$$

As we shall see in the next section, an applied forward bias $V = V_f$ increases the probability that a carrier can diffuse across the junction, by the factor exp (qV_f/kT). Thus the diffusion current under forward bias is given by its equilibrium value multiplied by exp (qV_f/kT); similarly, for reverse bias the diffusion current is the equilibrium value reduced by the same factor, with $V = -V_r$. Since the equilibrium diffusion current is equal in magnitude to $|I(\text{gen.})|$, the diffusion current with applied bias is simply $|I(\text{gen.})| \exp(qV_f/kT)$. The total current I is then the diffusion current minus the absolute value of the generation current, which we will now refer to as I_0

$$I = I_0(e^{qV/kT} - 1) \tag{1.20}$$

In Eq. (1.20) the applied voltage V can be positive or negative, $V = V_f$ or $V = -V_r$. When V is positive and greater than kT/q ($kT/q = 0.0259$ V at room temperature), the exponential term is much greater than unity. The current thus increases exponentially with forward bias. When V is negative (reverse bias), the exponential term approaches zero and the current is $-I_0$, which is in the n to p (negative) direction. This negative generation current is also called the reverse saturation current. Current flows relatively freely in the forward direction of the diode, but almost no current flows in the reverse direction.

From the discussion in the previous section we expect the minority carrier concentration on each side of a p-n junction to vary with the applied bias because of variations in the diffusion of carriers across the junction. The equilibrium ratio of hole concentrations on each side

$$p_p/p_n = e^{qV/kT} \tag{1.21}$$

Becomes with bias

$$p(-x_{p0})/p(x_{n0}) = e^{q(V_0 - V)/kT} \tag{1.22}$$

This equation is obtained by the same mathematics as Eq. (1.21), but with the altered barrier $V_0 - V$; it relates the steady state hole concentrations on the two sides of the transition region with either forward or reverse bias (V positive or negative). For low-level injection we can neglect changes in the majority carrier concentrations, which vary only slightly with bias compared with their equilibrium values. With this simplification we can write the ratio of Eq. (1.21) to Eq. (1.22) as

$$p(x_{n0})/p_n = e^{qV/kT} \tag{1.23}$$

With forward bias, Eq. (1.23) suggests a greatly increased minority carrier hole concentration at the edge of the transition region on the n (x_{n0}) than the case at equilibrium. Conversely the hole concentration $p(x_{n0})$ under reverse bias (V negative) is reduced below the equilibrium value p_n. The exponential increase of the hole concentration at x_{n0} with forward bias is an example of minority carrier injection. We can easily calculate the excess hole concentration Δp_n at the edge of the transition region by subtracting the equilibrium hole concentration from Eq. (1.23).

$$\Delta p_n = p(x_{n0}) - p_n = p_n(e^{qV/kT} - 1) \tag{1.24}$$

From our study of diffusion of excess carriers in last section, we expect that injection leading to a steady concentration of Δp_n excess holes at x_{n0} will produce a distribution of excess holes in the n material. As the holes diffuse deeper into the n region, they recombine with electrons in the n material, and the resulting excess hole distribution is obtained as a solution of the diffusion equation. If the n region is long compared with the hole diffusion length L_p, the solution is exponential. Similarly, the injected electrons in the p material diffuse and recombine, giving an exponential distribution of excess electrons. For convenience, let us define two new coordinates: distances measured in the x-direction in the n material from x_{n0} will be designated x_n; distances in the p material measured in the $-x$-direction with $-x_{p0}$ as the origin will be called x_{p0}. This convention will simplify the mathematics considerably. We can write the diffusion equation for each side of the junction and solve for the distributions of excess carriers assuming long n and p region

$$\delta_p(x_n) = \Delta p_n e^{-x_p/L_n} = p_n(e^{qV/kT} - 1)\, e^{-x_p/L_p} \tag{1.25a}$$

$$\delta_n(x_n) = \Delta p n_{np} e^{-x_n/L_p} = n_p(e^{qV/kT} - 1)\, e^{-x_n/L_n} \tag{1.25b}$$

The hole diffusion current at any point x_n in the n material can be calculated

$$I_p(x_n) = -qAD_p\delta_p(x_n)/dx_n = qAD_p/L_p\delta_p(x_n) \tag{1.26}$$

where A is the cross-sectional area of the junction. Thus the hole diffusion current at each position is proportional to the excess hole concentration at that point. The total hole current injected into the n material at the junction can be obtained simply by evaluating Eq. (1.26) at x_{n0}

$$I_p(x_n = 0) = qAD_p/L_p\Delta p_n = (qAD_p/L_p)p_n(e^{qV/kT} - 1) \tag{1.27}$$

The minus sign in Eq. (1.27) means that the electron current is opposite to the direction; that is, the true direction of I_n is in the $+x$-direction, adding to I_p in the total current. If we neglect recombination in the transition region, we can consider that each injected electron reaching $-x_{p0}$ must pass through x_{n0}. Thus the total diode current I at x_n can be calculated as the sum of $I_p(x_n = 0)$ and $-I_n(x_p = 0)$. If we take the direction as the reference direction for the total current I, we must use a minus sign with $I_n(x_p)$ to account for the fact that x_p is defined in the $-$x-direction

$$I = qA\left(\frac{D_p}{L_p}p_n + \frac{D_n}{L_n}n_p\right)(e^{qV-kT} - 1) = I_0(e^{qV-kT} - 1) \tag{1.28}$$

Eq. (1.27) is the diode equation having the same form as the qualitative relation Eq. (1.20). Nothing in the derivation excludes the possibility of the total current through the diode for either forward or reverse bias. We can calculate the current for reverse bias by letting $V = -V_r$.

Another simple and instructive way of calculating the total current is to consider the injected current as supplying the carriers for the excess distributions. For example, $I_p(x_n = 0)$

must supply enough holes per second to maintain the steady state exponential distribution Δp_n as the holes recombine. The total positive charge stored in the excess carrier distribution at any instant of time is

$$Q_p = qAL_p\Delta p_n \tag{1.29}$$

The average lifetime of a hole in the n-type material is τ_p. Thus on the average, this entire charge distribution recombines and must be replenished every second. The injected hole current (at $x_n = 0$) needed to maintain the distribution is simply the total charge divided by the average time of replacement

$$I_p(x_n = 0) = Q_p/\tau_p = qA\Delta p_n D_p/L_p \tag{1.30}$$

using

$$D_p/L_p = L_p/\tau_p$$

The result is the same as Eq. (1.26), which was calculated from the diffusion currents. Similarly, we can calculate the negative charge stored in the distribution and divide by τ_n to obtain the injected electron current in the p material. This method, called the charge control approximation, illustrates the important fact that the minority carrier current [for example, $I_p(x_n)$] decreases exponentially with distance into the neutral region. Therefore, several diffusion lengths are far away from the junction, most of the total current is carried by the majority carriers. We shall discuss this point in more detail in the following section.

In summary we can calculate the current at a p-n junction in two ways ① from the slopes of the excess minority carrier distributions at the two edges of the transition region and ② from the steady state charge stored in each distribution. We add the hole current injected into the n material $I_p(x_n = 0)$ to the electron current injected into the p material $I_n(x_p = 0)$, after including a minus sign with $I_n(x_p)$ to conform with the conventional definition of positive current in the + x-direction. We are able to add these two currents because of the assumption that no recombination takes place within the transition region. Thus we effectively have the total electron and hole current at one point in the device (x_{n0}). Since the total current must be constant throughout the device (despite variations in the current components), I as described by Eq. (1.28) is the total current at every position x in the diode.

One implication of Eq. (1.28) is that the total current at the junction is dominated by the injection of carriers from the more heavily doped side into the side with lesser doping. For example, if the p material is very heavily doped and the n region is lightly doped, the minority carrier concentration on the p side (n_p) is negligible compared with the minority carrier concentration on the n side (p_n). Thus the diode equation can be approximated by the injection of holes only, as in Eq. (1.27). This means that the charge stored in the minority carrier distributions is due mostly to holes on the n side. This structure is called a p^+-n junction, where the + superscript simply means heavy doping. Another characteristic of the p^+-n or n^+-p structure is that the transition region extends primarily into the lightly doped region. Having one side heavily doped is a useful arrangement for many practical devices, as we shall see in our discussions of switching diodes and transistors. This type of junction is common in devices which are fabricated by counterdoping. For example, an n-type Si sample with $N_d = 10^{14}\ cm^{-3}$ can be used as the substrate for an alloyed or diffused junction. If the doping of the p region is greater than $10^{19}\ cm^{-3}$ (typical of alloyed junctions), the structure is definitely p^+-n, with n_p more than five orders of magnitude smaller than p_n. Since this configuration is common in device technology, we shall return to it in much of the following discussion.

In this discussion of carrier injection and minority carrier distributions, we have primarily assumed forward bias. The distributions for reverse bias can be obtained from the same equations, if a negative value of V is introduced. For example, if $V = -V_r$ (p negatively biased with respect to n), we can approximate Eq. (1.24) as

$$\Delta p_n = p_n[e^{q(-V_r)/kT} - 1] \approx -p_n \text{ for } V_r \gg kT/q \tag{1.31}$$

and similarly $\Delta n_p \approx -n_p$.

Thus for a reverse bias of more than a few tenths of a volt, the minority carrier concentration at each edge of the transition region becomes essentially zero as the excess concentration approaches the negative of the equilibrium concentration. The excess minority carrier concentrations in the neutral regions are still given by Eq. (1.25), so that depletion of carriers below the equilibrium values extends approximately a diffusion length beyond each side of the transition region. This reverse-bias depletion of minority carriers can be thought of as minority carrier extraction, analogous to the injection of forward bias. Physically, extraction occurs because minority carriers at the edges of the depletion region are swept down the barrier at the junction to the other side and are not replaced by an opposing diffusion of carriers. For example, when holes at x_{n0} are swept across the junction to the p side by the ε field, a gradient in the hole distribution in the n material exists, and holes in the n region diffuse toward the junction. It is important to remember that although the reverse saturation current occurs at the junction by drift of carriers down the barrier, this current is fed from each side by diffusion toward the junction of minority carriers in the neutral regions. The rate of carrier drift across the junction (reverse saturation current) depends on the rate at which holes arrive at x_{n0} (and electrons at x_{p0}) by diffusion from the neutral material. These minority carriers are supplied by thermal generation, and we can show that the expression for the reverse saturation current, represents the rate at which carriers are generated thermally within a diffusion length of each side of the transition region.

New Words and Expressions

abrupt 突变的
acceptor 受体
bcc 体心立方
bipolar transistor 双极晶体管
conduction band 导带
conductivity 电导率
covalent bonding 共价键
decay 衰减
diamond lattice 金刚石晶格
donor 给体

energy levels 能级
exponential 指数的；幂数的
extrinsic 非本征的
fcc 面心立方
hole 空穴
impact ionization 碰撞电离
mass action law 质量作用定律
mean free time 平均自由时间
Miller indices 密勒指数
mobility 迁移率
recombination 复合
rectifier 整流器
resistivity 电阻率
scatter 散射
semiconductor 半导体
transition 跃迁

Translation

1. Insulators such as fused quartz and glass have very low conductivity.

翻译：像熔融石英和玻璃这样的绝缘体具有非常低的导电性。

2. This sensitivity in conductivity makes semiconductor one of the most important materials for electronic applications.

翻译：这种导电性的敏感度使半导体成为电子应用中最重要的材料之一。

3. The element semiconductors, those composed of single species of atoms, such as silicon (Si) and germanium (Ge), can be found in Column Ⅳ.

翻译：如硅（Si）和锗（Ge）这些由单个原子组成的元素半导体，可以在第四族找到。

4. However, germanium is proved unsuitable in many applications because germanium devices exhibited high leakage currents at only moderately elevated temperatures.

翻译：然而，锗在许多应用中被证明是不合适的，因为锗器件在适度升高的温度下表现出很高的泄漏电流。

5. In addition, germanium oxide is water soluble and unsuited for device fabrication.

翻译：另外，氧化锗是水溶性的，不适合制作器件。

6. At present, silicon is one of the most studied elements in the periodic table; and silicon technology is by far the most advanced among all semiconductor technologies.

翻译：目前，硅是元素周期表中被研究最多的元素之一，而硅技术是迄今为止所有半导体技术中最先进的。

7. The semiconductor materials we will study are single crystals, that is, the atoms are arranged in a three-dimensional periodic fashion.

翻译：我们要研究的半导体材料是单晶半导体，也就是原子以三维周期的方式排列的半导体。

8. Therefore, the crystal properties along different planes are different, and the electrical and other device characteristics are dependent on the crystal orientation.

翻译：因此，沿不同平面晶体的性质是不同的，并且电学和其他器件特性取决于晶体的取向。

9. At low temperatures, the electrons are bound in their respective tetrahedron lattice; consequently, they are not available for conduction.

翻译：在低温下，电子被束缚在各自的四面体晶格中，因此，它们不可用于传导。

10. Diffusion current results from the random thermal motion of carriers in a

concentration gradient.

翻译：扩散电流是载流子在浓度梯度上随机热运动的结果。

11. We can inject carriers by using various methods including optical excitation and forward biasing a p-n junction.

翻译：我们可以用各种方法注入载流子，包括光激发和给 PN 结加正向偏压。

12. Recombination phenomena can be classified as direct and indirect processes.

翻译：复合现象可分为直接过程和间接过程。

Unit 2
Semiconductor Devices

2.1 Bipolar Junction Transistors

The bipolar junction transistor (BJT) was invented in January 1948, a few weeks after the invention of the first transistor (the point-contact transistor), and months before the point-contact device was announced. The reason for the invention of the BJT is a prime object lesson in engineering practice. The point-contact transistor was fabricated by primitive and "artistic" methods, and its detailed structure and operation remain unclear to this very day. Even worse, it posed a complex three-dimensional problem. The inventor of the BJT, William Shockley, was at the time the supervisor of the two inventors (or "discoverers") of the point-contact transistor, John Bardeen and Walter Brattain. He was endeavoring, with sound engineering instinct, to conceive a one-dimensional analog of the point-contact device as a way of opening the door to analytical investigation of its operation. Suddenly he realized that the structure he was postulating was itself a transistor, and onto that lent itself to analysis at a basic level! It became the bipolar junction transistor. Reduced to practice in 1951, it had for all practical purposes shouldered the point-contact device aside by the mid-1950s. A genuine understanding of the point-contact transistor will probably never be achieved, because it is inferior to competing devices in so many respects that it no longer arouses even academic interest. Nonetheless, its place in history is assured as the first practical amplifying device in the era of solid-state electronics.

Early in the 1950s, Shockley proposed the adjective bipolar to describe his newly conceived device because both carrier types play important roles in its operation; by contrast, he described devices in the field-effect family (which he also launched insofar as practical embodiments were concerned) as unipolar, because their operation depends primarily on a single carrier type. In more recent times, the need to distinguish among the proliferating varieties of transistors has caused the term bipolar junction transistor to come into common use, and resulting acronym BJT to come into even more common use. This kind of distinction is important. Shockley's first field-effect device was a "unipolar junction transistor", but it is usually designated the junction field-effect transistor. The term "unipolar" is not widely used.

The unfolding story of solid-state electronics can be told rather completely in terms of evolving fabrication technology, constantly expanding the number of options available to the device and integrated-circuit designer. It was for technological reasons that an early and important kind of BJT was a germanium PNP device. The term PNP labels the conductivity types of the three regions within a BJT, regions separated by two PN junctions. In later years, and again partly for technological reasons, the dominant BJT was a silicon NPN device. In integrated circuits today, the combination of silicon NPN and PNP devices is a growing practice because the resulting complementary circuits have important power-dissipation and performance advantages. For convenience and consistency, however, and because of its continuing importance, the silicon NPN BJT will be the vehicle for this chapter.

2. 1. 1 Structure and Technology

The essential structure of a BJT is represented in Figure 2-1 (a). The very earliest devices had structures literally of this kind. Two closely spaced junctions were created by crystal-growth methods, and a " bar " or parallelepiped was then cut out of the germanium crystal. Electrical leads were attached to it and the result was a BJT. For reasons that will be explained shortly, these electrical terminals are given the names, respectively from left to right, emitter, base, and collector. These names were chosen with an eye to distinctive initial letters, which are displayed in Figure 2-1 (a) in association with the three terminals. The shaded regions in Figure 2-1 (a) represent the space-charge regions (or depletion regions) of a pair of NP junctions. The boundaries of these regions are emphasized because conditions there assume far greater importance than conditions at the metallurgical junctions (which are not even represented in the drawing).

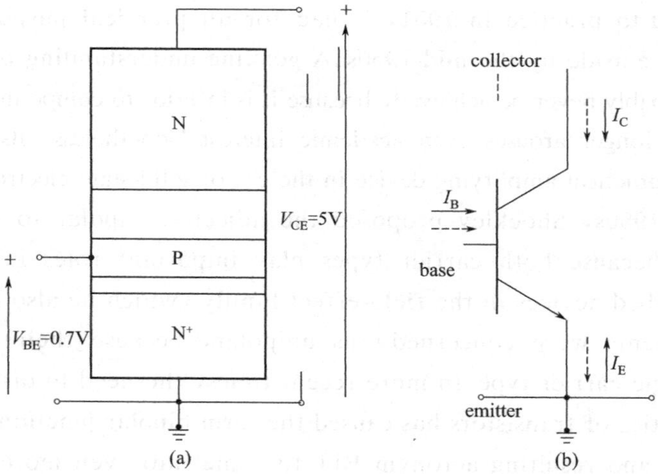

Figure 2-1 A representation of the NPN bipolar junction transistor (BJT)

Having given names and single-letter labels to the three terminals of the BJT, we have effectively provided names and labels to the three regions of the BJT as well, because each region is associated uniquely with one terminal. The matter of junction, identification is almost as simple. The two junctions are, respectively, the emitter-base junction and the collector-base junction. However, without any ambiguity we can (and usually do) refer to them simply as the emitter junction and the collector junction.

The corresponding BJT symbol, Figure 2-1 (b), was adopted very early. Once again an arrowhead is used to point from a P-type region to an N-type region, the direction of the "easy" conduction of conventional current through a PN junction. But only one junction—the emitter junction is so labeled. (This is done for identification.) Once the emitter junction has been so identified and labeled, the polarity of the collector junction becomes evident.

In 1960 the BJT took on a structure that it has retained to the present day in a wide range of applications, the structure is shown in Figure 2-2. The orientation of this diagram has been chosen to correspond to that of the BJT in Figure 2-1 (a). It is evident that this structure bears little resemblance to the prototypical "bar" structure. Nonetheless, in a large number of devices having the newer structure, the currents through the junctions retain an essentially one-dimensional character, with important current fractions and current components being normal to the junctions. Consequently, a good approximate treatment of such devices can be achieved by calculating currents in the region emphasized by dotted lines in Figure 2-2 and then multiplying by an appropriate junction area to determine total current through the junction. Thus Shockley's original one-dimensional intent is still being honored.

Figure 2-2 specifically represents the structure of a discrete or separately packaged BJT. It is evident from the dimensions shown that the active portion of the device is confined to a small portion of the overall crystal volume. The bulk of the silicon crystal, in fact, is simply a mechanical support for the active portion. Any electrical resistance to the passage of current through this thick region is a parasitic feature of the device that one would prefer to avoid. Most BJTs today are incorporated into an integrated circuit (the "microchip" of the layman), defined as a useful combination of devices fabricated in and in some cases on a semiconductor single crystal. As a result, the parasitic feature of having current pass through a thick mechanical support is partly eliminated. All of the currents (in virtually all integrated circuits) are confined to a very thin region near one face of a semiconductor crystal.

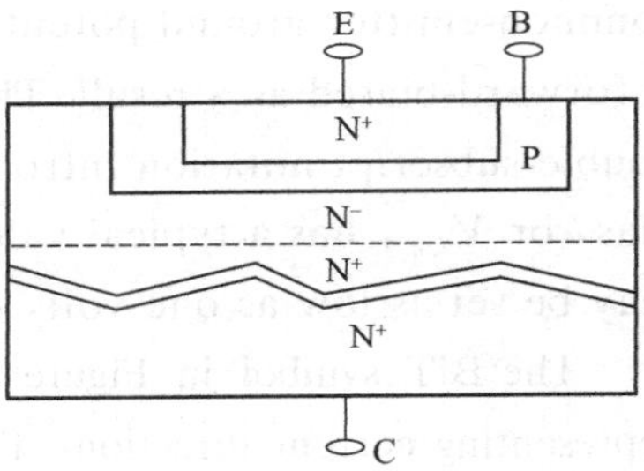

Figure 2-2 A BJT structure, employing the epitaxial growth of silicon

The portions of the BJT structure shown in Figure 2-2 that are near the top and the bottom of the diagram also are to some degree "parasitic". (For a more usual orientation of the diagram with the silicon surface horizontal, these regions would be described as lateral portions of the structure.) It is a curious fact that advancing technology is now generating structures that eliminate the parasitic lateral regions, thus increasingly resembling the original bar structure! These BJTs are not freestanding parallelepipeds, but rather are defined by parallel grooves cut into a silicon crystal, and then subsequently filled by an insulating material such as silicon dioxide (SiO_2). The individual BJTs are interconnected to perform useful circuit functions by means of a variety of methods; the result is an integrated circuit that may incorporate over 100000 devices all fabricated simultaneously in the same silicon crystal.

2. 1. 2 Biases and Terminal Currents

An input port or an output port in the circuit sense consists of a pair of terminals. Since the BJT is a three-terminal device, one of the three terminals is permitted to be common to the input and output ports, and the other two terminals are each uniquely associated with each port. For reasons that we shall examine, when the base and emitter terminals are chosen for the input port with the collector and emitter terminals taken for the output port, the BJT can exhibit both current gain and voltage gain. This useful combination of properties has made the common-emitter configuration the most widely used of the several possibilities, a term acknowledging that the emitter terminal is common to the input and output ports. The adjective "grounded-emitter" is also sometimes applied to this connection.

Figure 2-1 (a) shows the physical representation of the BJT in its common-emitter orientation, and also shows typical voltage polarities and magnitudes for the case of a silicon device. A positive voltage is applied at the input port (with reference to the common-emitter ground potential shown here), and it is evident that the emitter junction is forward-biased as a result. The junction voltage is near 0. 7 V. Or, continuing to use the double-subscript notation introduced in Chapter 1, we have $V_{BE} \approx 0.7$ V. Collector-emitter bias, or V_{CE}, has a typical value of 5 V in wide ranges of circuits, but for some purposes may be set as low as one volt, or as high as many hundreds of volts.

The BJT symbol in Figure 2-1 (b) has associated with each terminal a pair of arrows representing current directions. The solid arrows show the directions of conventionally positive terminal current-always in ward. The dashed arrows show the actual directions of positive conventional current at each terminal for the bias conditions shown in Figure 2-1 (a). Note that in the cases of base and collector currents, the two most important to the common-emitter problem, the two current conventions are in agreement with respect to direction, a matter of significant convenience.

Operation of the BJT can be approached by examining the properties of its two junctions individually, considering them to be isolated. As we saw in Chapter 1, this is the term that describes a junction whose end regions are extensive in the x direction (since we continue to consider one-dimensional structures only). As a result, the carrier-density disturbances that accompany junction biasing have space enough to "fade away" before they reach the inevitable contacts. Thus the inherent properties of the junction under study will be seen. And since we are going to assume step junctions only, these inherent junction properties subsume those of both of its uniformly doped regions—properties such as absolute doping and carrier lifetime. Then we shall combine the junctions in order to examine their interaction. In a BJT, this interaction is crucial; there is no way to simulate BJT behavior using isolated junctions.

Figure 2-3 (a) shows the BJT structure. The dimensions of the various regions are shown with qualitative realism. The emitter-junction space-charge layer is typically much thinner than that of the collector junction, for two reasons: First, the mean doping of the emitter and base regions (a geometric mean is appropriate here because their net-doping values have a large ratio) is typically almost two orders of magnitude larger than the mean doping of the base and collector regions; and second, the emitter junction is usually forward-biased, while the collector junction is usually reverse-biased. The thickness of the latter's space-charge region is designated as X. The thickness of the base region X_B is typically a few tenths of a micrometer in a modern transistor, and is a critical dimension.

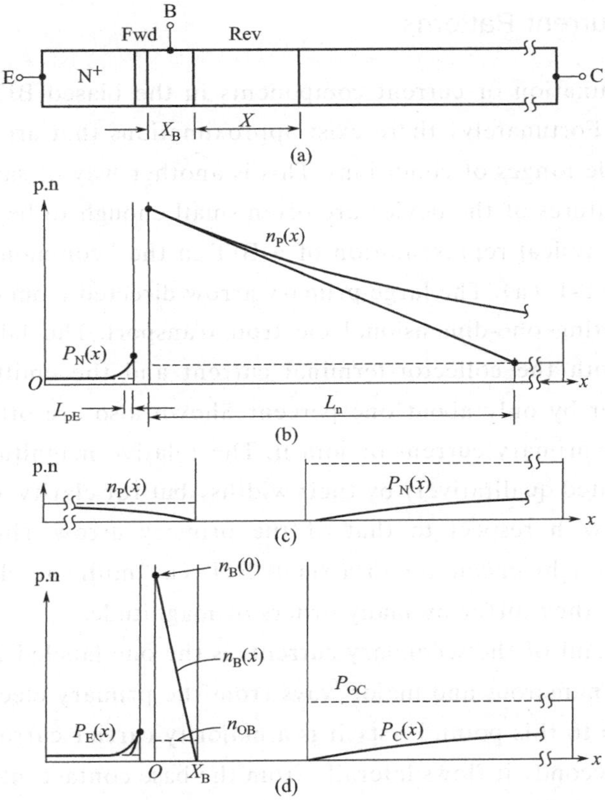

Figure 2-3 Minority-carrier profiles in a normally biased BJT

Since the late 1950s, various techniques involving the high-temperature solid-phase diffusion of impurities have been extensively employed in BJT fabrication. In particular it has been emitter regions and base regions that have been formed by these methods, yielding BJT cross sections well represented by Figure 2-2. The impurity profiles that result from impurity-diffusion procedures are not the constant-density profiles incorporated in an ideal-step-junction description. Instead they approximate the Gaussian function and the error-function complement as described in Chapter 1. As a result, the net-doping-density profiles for both the emitter and base regions exhibit maxima that are skewed toward the surface of the device.

Fortunately for the student and engineer, it turns out that these rather major departures from the ideal structures defined in Figure 2-3 are less consequential than one might normally assume. For a combination of reasons, calculations based upon idealized structures can provide meaningful descriptions of device properties and worthwhile guides for design modifications. Therefore, in what follows, we shall continue to adhere to the assumptions of step junctions and uniformly doped regions. It is a curious fact, that modem technology is moving ever closer to the step-junction ideal employed in the simplest analyses, with uniform doping in the emitter and collector regions, and sometimes in the base region as well.

2. 1. 3 Internal Current Patterns

A detailed examination of current components in the biased BJT reveals a relatively complicated picture. Fortunately, there exist approximations that are simple and yet quite accurate through wide ranges of conditions. This is another way of saying that some of the most complicated features of the device are often small enough to be neglected safely.

Figure 2-4 is a physical representation of a BJT in the "common-emitter orientation" shown first in Figure 2-1 (a). The large primary arrow directed from collector to emitter is identically, representing one-dimensional electron transport. The labels I_C and I_E show that it constitutes both the collector-terminal current and the emitter-terminal current, which typically differ by only about one percent. Shown also are other current "strands" that depart from the primary current or join it. The relative magnitudes of the secondary currents are represented qualitatively by their widths, but for clarity all of these widths are grossly exaggerated with respect to that of the primary arrow. The circled values give typical currents one might encounter in a small BJT (a "milliwatt device") under bias, and it is evident that they differ by many orders of magnitude.

The most important of the secondary currents is the one labeled I_B, the base-terminal current. It differs in numerous and major ways from the primary electron current that has occupied our lives up to this point. First, it is a majority-carrier current, or a hole current in the NPN device. Second, it flows laterally from the base contact into the active region of the BJT. Refer back to Figure 2-2, which shows the commonplace (diffused) BJT in a relatively realistic cross section. The portion of the base region contiguous with (or "under") the emitter region is the active base region, sometimes also termed the intrinsic base region. (This use of the term does not have doping connotations.) The outer portions of the base region, in analogous fashion, are sometimes designated by the adjective extrinsic. It is there that the base contact is made in this kind of BJT. (Base contacts, by the way, can be and often are made on both sides of the emitter.) The third major distinction of the base current from the primary electron current thus becomes evident. The current of holes traverses a relatively long path within the base region. In the example we are considering it could be several tens of micrometers, as compared to the less-than-one-micrometer base thickness traversed by the electrons.

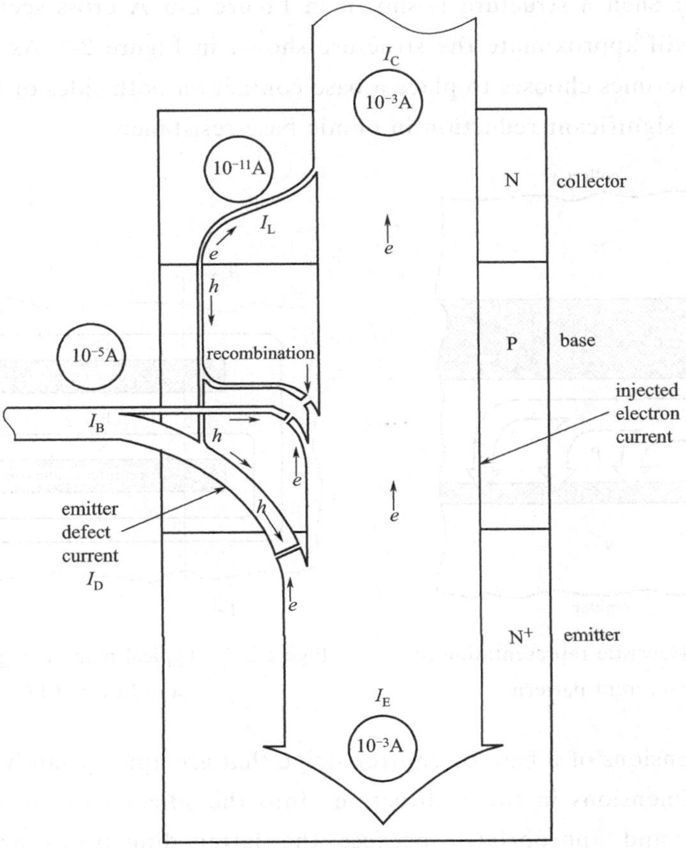

Figure 2-4 Representation of electron and hole (e and h) current component in BJT under bias

Once in the intrinsic base region, most of the holes undergo a change in direction and flow into the transition region of the emitter junction. This obviously involves a complex current pattern. It is not a "one-piece" current like that shown for simplicity and convenience in Figure 2-4, but rather is somewhat like the pattern shown heuristically in Figure 2-5. In spite of the relatively small size of the base current, it is of extreme importance because, as we shall see, it is the control current that is supplied to the input port in the common-emitter BJT. The perturbing effects of the complicated base-current pattern on, for example, the electrostatic-potential pattern in the base region, are not serious so long as the base current is small. In the typical device we are considering base current amounts roughly to one percent of the electron current, which does qualify as "small".

The resistance encountered by the base current on its lateral path from the (ideal, let us say) base contact to the active base region, and in the active base region as well, plays a key part in fixing device performance. This ohmic base resistance is a parasitic element to be minimized. Its value is typically in the range of 50 to 100 Ω. It is in order to diminish base resistance, that the emitter is often given the (plan-view or top-view) shape of an elongated rectangle or stripe, with the base contact, also, stripe like, in close proximity to the emitter stripe. Such a structure is shown in Figure 2-6. A cross section of this device taken at A—A will approximate the structure shown in Figure 2-2. As noted earlier, the BJT designer sometimes chooses to place a base contact on both sides of the emitter stripe; doing so makes a significant reduction in ohmic base resistance.

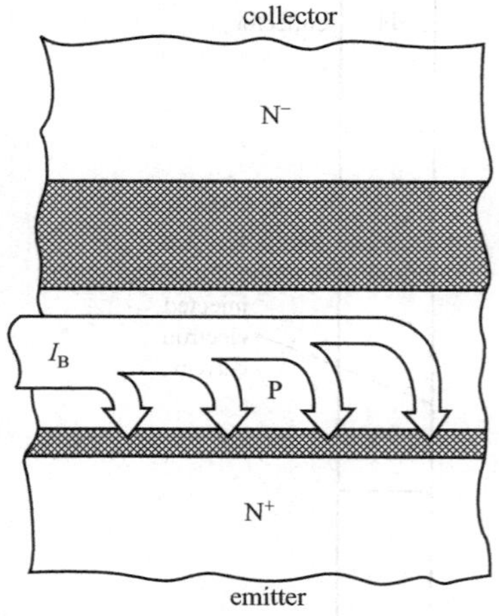

Figure 2-5 Heuristic representation of base-current pattern

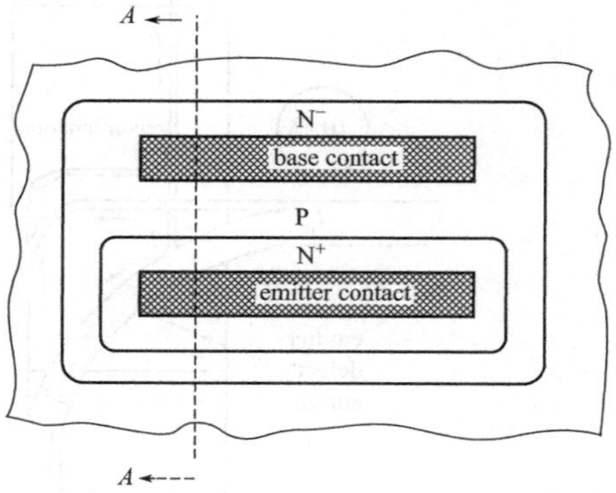

Figure 2-6 Typical plan-view geometry for a milliwatt BJT

It is the dimensions of a base or emitter stripe that are appropriately designated length and width. For dimensions in the x direction, into the silicon, the term thickness is far more descriptive and appropriate, because the latter dimensions are typically much smaller. More important, clarity is served by consistency in these matters. Curiously, even engineers, who are supposed to have above-average skill in spatial visualization, sometimes seem to have difficulty in understanding and verbalizing the three-dimensional aspects of the BJT. One too frequently hears and reads tangled syntax such as "this transistor has a very thin base width" instead of the simple declaration that the transistor has a thin base region.

Most of the holes shown entering the emitter junction in Figure 2-5 pass through it and recombine in the low-lifetime N^+ emitter region, so long as the BJT is not being operated with extremely low current levels. These holes are identically those represented by the profile labeled $P_E(x)$ that is shown at the left in Figure 2-3 (d). In Figure 2-6, the recombination is symbolized by the gap in the I_B arrow inside the emitter region, with holes approaching it from the base side and the recombining electrons being supplied within the emitter.

2.1.4 Parasitic Internal Currents

The remaining two currents are parasitic in the sense that they vanish in the ideal BJT, and nearly vanish in the silicon BJT. First, a small current is shown in Figure 2-4 as splitting from I_B to a recombination gap within the base region. A small but finite portion of the flood of electrons diffusing from emitter to collector fails to reach the collector, recombining in transit, and it is this occurrence that the second gap symbolizes. In the early BJTs, most of the base current was consumed by this mechanism because the base thickness X_B amounted to tens of micrometers and base-region lifetime values τ_B were sometimes low. As a result, the crucial electron profile in the base region departed significantly from the modem-device linear profile shown in Figure 2-3 (d). The base-transport efficiency γ_B took on considerable importance in those days. (Today it is so close to unity that it is rarely mentioned.) The factor γ_B can be defined as the ratio of electron current departing from the active base region at its right-hand surface to the electron current entering the active base region at its left-hand surface.

Returning now to the modem transistor, we stress that I_B is very nearly equal to the current of holes being injected into the emitter. This brings us to a second efficiency factor, γ_E, one that is of great importance in the modem BJT. It can be defined as

$$\gamma_E = (I_{nE}/I_E) = (I_E - I_{pE})/I_E = 1 - (I_{pE}/I_E) \tag{2.1}$$

where I_{nE} and I_{pE} are, respectively, the electron current being injected into the base region by the emitter junction and the hole current being injected into the emitter region by the same junction. The total emitter-junction current is inevitably equal to the emitter-terminal current I_E. In a "perfect" emitter, all of the emitter current would consist of electrons injected into the base region. From this point of view I_{pE} (the current of holes injected into the emitter region) is a defect current, a description sometimes used. The adjective "perfect" was placed in quotation marks, however, because the injected current of holes plays the critical role of the control current, and as such is a vital factor in BJT design and operation, rather than a factor to be eliminated as the potentially misleading term "defect current" may suggest.

This brings us to the last and smallest current, namely the leakage current I_L arising at the reverse-biased collector junction. It provides one of the sharpest contrasts between the modern silicon transistor and the early germanium transistor. As we noted in Chapter 1, the behavior of a reverse-biased silicon junction fits simple theory quite poorly, while the germanium case fits it very well. At the same time, the silicon-junction features that make it nonideal from a theoretical point of view serve to create near-ideal BJT properties! These details are worth examining because of the insight they provide into certain BJT subtleties. A lesser reason is that part of the older literature on the BJT and its applications dwells on the nonideal BJT properties that arise from the "ideal" germanium-junction behavior, and this literature is puzzling to a newcomer without a measure of appreciation of the responsible phenomena.

We saw in Chapter 1 that the dominant reverse-current mechanism in a silicon junction is the excess of generation over recombination in the space-charge region, where reverse bias has caused a depression of both carrier densities below their equilibrium values. (In such conditions, generation exceeds recombination, just as the reverse is true in the presence of excess carriers.) The carriers so generated are separated by the electric field and are swept to the regions where they become majority carriers. A representative value for this leakage current is of the order of ten picoamperes. The description just given shows that it consists equally of holes and electrons. The electrons become a negligible part of the collector current, departing the BJT through the collector terminal. The holes entering the base region mingle with those coming from the base contact. As a result, they participate in emitter-region recombination and, to a much lesser degree, base-region recombination.

Conditions in the old germanium BJT were quite different. First, the diffusion of minority carriers to the reverse-biased collector junction provided most of the collector leakage current. Because the collector region is more lightly doped than the base region by one to two orders of magnitude, minority holes in the collector region dominate this diffusion mechanism. This density is roughly four million times longer in germanium than in silicon for given net-doping values, because the values of n_i^2 stand in about that ratio for the two materials, and their mobility ratios introduce another factor of about three. The resulting current of holes into the base region might be of the order of 0.1 microampere, amounting to one percent of the 10 μA base current cited in the example of Figure 2-4. When I_B is further diminished (and in particular, when it is reduced to zero), this leakage current is like an irreducible "parasitic base current". Further, this parasitic current is then multiplied by the gain mechanism of the BJT, typically by 100, with very undesirable effects. The details of this topic are best treated after we have introduced and described the gain mechanism itself, which we will do shortly. And after that, we shall point out why the silicon BJT does not multiply its collector-leakage current (already three orders of magnitude smaller) by its current-gain factor.

2.1.5 Common-Emitter Current Gain

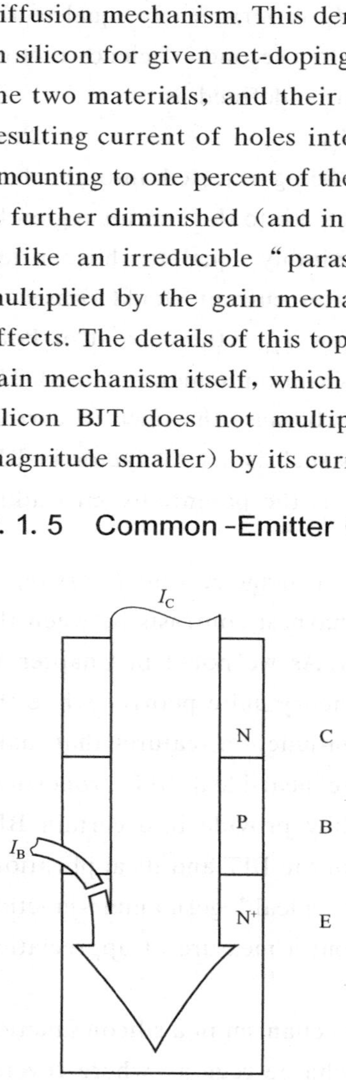

Figure 2-7 The terminal base and collector currents under typical conditions in a BJT

In the silicon BJT we may safely neglect for a wide range of purposes the two parasitic internal currents, namely, the base-recombination current and the collector-leakage current. Thus the admittedly complex pattern of Figure 2-4 reduces to the much simpler pattern of Figure 2-7. Even here, however, there remains the complexity of the base current's two-dimensional pattern, a complication that may be neglected for a small ratio of base current to electron current, as was noted earlier.

Return briefly to Figure 2-3 (d). Focus on the carrier-density profiles in the emitter and base regions, and ignore that in the collector region because it affects only collector-leakage current, now dropped from consideration. Elementary diffusion theory permits us to write the current of electrons in the base region. This constitutes the collector current I_C, which is the output current in the common-emitter situation under study. Having already established that it is a positive terminal current, let us employ absolute-value signs on the diffusion-current expression. Given the emitter-junction area A_E

$$I_C = |A_{Eq} D_{nB} (dn_B/dx)| \tag{2.2}$$

Where the gradient can be treated as constant throughout the base region is given simply by $n_B(0)/X_B$. IT is important to stress once more at this point that X_B is taken to be the distance from boundary to boundary, and not from (metallurgical) junction to junction. It is clearly that the base region as defined by the respective depletion-layer boundaries enters into the gradient calculation. Thus X_B is the effective base thickness. To repeat, it is the thickness dimension of the active (or intrinsic) base region, whose lateral dimensions are fixed by emitter plan-view shape.

Employing the law of the junction now yields $n_B(0)$ as a function of applied voltage, in turn giving us I_C as a function of base-emitter voltage. The voltage $-V_{NP}$ in our law-of-the-junction formulation is equivalent to $-V_{EB}$, and hence to $+V_{EB}$. Therefore

$$I_C = (A_{Eq} D_{nB} n_{OB}/X_B)\exp(qV_{BE}/kT) \tag{2.3}$$

Identical principles permit us to write the current of holes diffusing into the emitter region, except that the critical length is L_{PE} rather than a device dimension. Accordingly,

$$I_B \approx |I_D| = |-A_{Eq} D_{PE} (dp_E/dx)| \tag{2.4}$$

From which

$$I_B = (A_{Eq} D_{PEPOE}/L_{PE})\exp(qV_{BE}/kT) \tag{2.5}$$

It is of the greatest importance that the same voltage V_{BE} enters into both the input and output currents. Because this is true, we can define a common-emitter static (or dc, or steady-state) current gain that is a function of structural constants in the device.

$$\beta = (I_C/I_B) \tag{2.6}$$

Making use of Eq. (2.3) and Eq. (2.5) give us

$$\beta = \frac{D_{nBnOB}/X_B}{D_{PE} P_{OE}/L_{PE}} \tag{2.7}$$

Thus from Eq. (2.6) we have

$$I_C = \beta I_B \tag{2.8}$$

Where β from Eq. (2.7) the device designer and fabricator can adjust through wide ranges. Because the output current I_C is proportional to the input current I_B, the common-emitter BJT is a linear current amplifier.

The range of common-emitter current-gain values, or β values, found in practical BJTs, extends through about three orders of magnitude. In power BJTs (ranging up to hundreds of watts), a combination of factors forces the designer to choose large values of X_B—several micrometers. As a result, such devices have low β—typically about ten. At the other extreme are superheat transistors, such as those at the input stages of operational amplifiers. Here the base regions are made so thin that β approaches 10000! The price that must be paid for this kind of gain performance is a drastic loss of operating-voltage range. As the collector space-charge layer expands under its normal reverse bias, its encroachment in an ultrathin base region can permit it to reach the emitter-junction space-charge region. Under these conditions, electrons are injected directly into the collector junction so that current increases precipitously and with no further control being exerted by the input current I_B. This condition is identically the punch through condition treated in Chapter 1.

In the era of germanium grown-junction and alloyed-junction BJTs, a large-signal model of fundamental importance was advanced by Ebers and Moll. In spite of extensive BJT structural evolution since that time, their model remains valid. Its primary use is for static modeling. The original paper extended the model to deal with large-signal, time-varying problems as well. But their dynamic model did not find wide acceptance, because it introduced for the extension somewhat awkward frequency-dependent current generators. For static modeling, however, the Ebers-Moll model is compact and economical. Furthermore, it is directly rooted in device physics. The physical properties stressed originally were common-base current gain, α_F and α_R, and the saturation currents exhibited by the two junctions in their BJT environment. The latter choice was a logical one in the germanium era, because as we have seen, germanium junctions are dominated by diffusion current in reverse bias and saturate cheaply. While the same approach still worked for silicon BJTs, a measure of physical relevance was lost. We now describe how that physical relevance was regained.

2. 1. 6 Gummel -Poon Reformulation

In 1970, a significant of the Ebers-Moll model was contributed by Gummel and Poon. It avoided the use of junction saturation currents, and common-base current gains, using instead α_F and α_R and the intercept current I_s that Gummel had previously introduced. The reformulated Ebers-Moll model contributed by Gummel and Poon has become known in its rudimentary or unelaborated embodiment as the transport form of the Ebers-Moll model. In addition to using device properties relevant to modem silicon transistors, it is appreciably more transparent than the model in original form. For that reason, we shall develop the transport from of the model equations first and move to the original form later, showing how the two are related.

While the transport-form reformulation was a significant contribution, the Gummel-Poon paper went well beyond it. Their primary interests were ① dynamic modeling for ② computer analysis of BJTs, ③ incorporating realistic properties. Setting aside idealized devices, they wanted their model to deal with such items as emitter crowding, the Early effect, the Sah-Noyce-Shockley effect (space-charge-region recombination), the Webster-Rittner effect, and the Kirk effect.

Their approach to this challenge employed a variation and extension of charge-control techniques. The original methods, described and applied to the isolated junction in Chapter 1, had been used extensively for the dynamic modeling of semiconductor devices. They were applied to the BJT by Beacfoy and Sparkes, with a result that is treated at some length in Chapter 1. Gummel himself contributed a different kind of charge-control model and applied it to the BJT in an earlier 1970 paper. The essence of this charge-control model "of the second kind" is illustrated in Section 2. 1. 7. The further extension contained in the Gummel-Poon paper used a "phenomenological" or semiempirical approach to the complicated problem of the real BJT. That is, their model does not concern itself with the details of carrier distributions and physical mechanisms responsible for charge storage. For example, in the case of the Kirk effect it simply associates a certain packet of charge with the collector junction, rather than becoming involved with information such as that in figure (which, to be sure, was not available at that time). Their computer-oriented model employed twenty-one parameters that were to be assigned and then empirically adjusted by comparing model predictions to experimental data. It has been extensively used in the years since its introduction. The book by Getreu gives a good account of related modeling developments in the 1970s.

Finally, significant success was achieved during the late 1970s and early 1980s in generating a fully analytic high-level model for the BJT. It can be regarded as a high-level version of the Ebers-Moll model. It combines previously separate analyses of high-level junction theory and ambipolar diffusion in the BJT context. Further, it yields accurate closed-form descriptions of certain effects (especially the Kirk effect) that previously had only been treated approximately, empirically, or numerically.

2. 1. 7 Assumptions and Problem Definition

A highly idealized device will be specified for present purposes. To leave no doubt about the device being analyzed, we offer a set of assumptions more detailed and explicit than those of Ebers and Moll:

① The emitter, base, and collector regions individually are uniformly doped. This requires that the junctions are step junctions.

② In the base and collector regions the doping is extrinsic but nondegenerate.

③ If the emitter-region doping is degenerate, the region can nonetheless be described by the conventional semiconductor equations by employing empirically adjusted values of quantities such as diffusivity and intrinsic density.

④ In the base and collector regions, carrier lifetime is high.

⑤ Injected-carrier current densities are small.

⑥ The I-V characteristic of each junction is of the form

$$I = -I_{KO}[\exp(-qV_{NP}/kT) - 1] \quad (2.9)$$

where I_{KO} is a constant and is valid for that junction in the presence of the opposite junction when the opposite junction is open-circuited. (Gummel and Poon chose to employ a short-circuited condition on the opposite junction, which requires a different coefficient in the equation for the junction I-V characteristic.)

⑦ The positional dependence of the junction boundaries upon bias voltage may be neglected.

⑧ Breakdown phenomena may be neglected.

⑨ The problem is one-dimensional.

⑩ The ohmic contacts are ideal and of the high-recombination-velocity variety.

⑪ The emitter and collector regions are thick, or extensive in the x direction, relative to their respective minority-carrier diffusion lengths.

⑫ The current gains α_F and α_R are independent of current and voltage.

These assumptions constitute a measure of "overkill" intended to simplify and clarify the model's insights and construction. The Ebers-Moll model, in fact, achieved remarkable generality. In an appendix to the original paper, the authors showed that the model was practically geometry-independent. Also, we saw in Section 2. 1. 2 that base-region doping non-uniformities has a smaller-than-expected impact upon low-level BJT properties. It was pointed out that junctions departing significantly from step-junction doping profiles exhibit properties with great qualitative similarity to those of step junctions.

The aim of the present exercise is to write expressions for the currents at two of the BJT terminals in terms of the voltages at those terminals, with the third terminal being a common reference for defining the voltages. If current at the third terminal is of interest, it can be readily determined from the other two currents by applying Kirchhoff's current law. To define the terminal currents and voltages, it is especially convenient to use the common-base configuration; each voltage is thus applied directly to one junction only. For forward bias, the respective voltages V_{EB} and V_{CB} must obviously be negative. Ebers and Moll chose to open-circuit one junction while biasing the other; on the other hand, Gummel and Poon, as noted earlier, chose the short-circuit option. This was the key step in achieving a more economical and transparent model.

2. 2 The MOSFET

A thin and tightly adherent layer of glass, mainly SiO_2, can be formed on the surface of a single-crystal silicon sample by heating it in an oxidizing atmosphere. The most common choices are oxygen and steam. Starting in the late 1950s, these procedures were moved along vigorously in the hope that they might provide at least a partial solution to the vexing problem of stabilizing surface conditions on silicon devices. Ultimately they permitted not only stabilization but also control of surface conditions; to accomplish this, however, enormous investments of time and resources were required to achieve the necessary understanding of the oxide-silicon system, and especially of the oxide-silicon interface.

At a relatively early stage, these procedures benefited the BJT, because its crucial phenomena occur mainly inside the silicon crystal. Further, even a limited grasp of oxide technology opened a whole new approach to BJT fabrication. But then came the MOSFET, which we shall describe in detail. By contrast with the BJT, the decisive phenomenon in the MOSFET occurs at the oxide-silicon interface. Consequently, the art and science of controlling interfacial properties required an order-of-magnitude improvement to make the MOSFET a truly practical device, a process that took approximately an additional ten years. The central portion of the MOSFET is an MOS capacitor, a structure that received intensive study in its own right. When isolated, it is a one-dimensional "sandwich" consisting of the oxide with silicon on one side, and with an adherent field-plate electrode on the opposite side that is capable of manipulating potential at the oxide-silicon interface.

2. 2. 1 Basic MOSFET Theory

As we saw in Section 2. 1, the essence of BJT operation is applying a voltage to modulate the potential barrier in a PN junction. The modulated current of carriers spilling over the barrier as a result becomes an output current, and these combined phenomena are responsible for the remarkable gain properties of this important bipolar device. The operating principle in a field-effect transistor, or FET, is quite different. A FET is essentially a voltage-controlled resistor. Hence it is "unipolar" in the sense that carriers that are in the majority in the region providing the resistance are dominate in determining device properties. As the result of a voltage applied from one end of the resistor to the other, these majority carriers flow ohmically in a thin region known as the channel. The carrier population in the channel is modulated by a separate control voltage that generates a population of carriers in a modulated electric field modulated that is substantially perpendicular to the thin channel. With the change of carrier number, the channel resistance also changes. The control electrode is known as the gate electrode.

There are several FET families, each with many members, and in each family there are differences of detail in how the control voltage interacts with the channel. Nonetheless, all of the FETs have similar gain properties, which are substantially inferior to those of the BJT through a wide current range. But, we should add quickly, the various families have offsetting advantages. In one case, that of the MOSFET, the sum of these advantages is so imposing that MOSFETs are the dominant solid-state devices in the world today.

2. 2. 2 Field-Effect Transistors

Well before the present solid-state-device era, intuitive thoughts were advanced on how to realize a solid-state amplifier; the structures proposed were essentially field-effect transistors. The serious postwar effort that led to the transistor also began with field-effect transistors. While that path was being explored, an unintended experimental event led to the discovery of the point-contact transistor, which found enough applications to be characterized as the first practical solid-state amplifier. But the point-contact transistor had serious shortcomings (and operating principles that remain obscure to this day). Consequently it was supplanted by the BJT within just a few years. The events connected with these two devices (occurring from the mid-1940s to the mid-1950s) display a fascinating blend of scientific insight, art, engineering intuition, and happenstance.

A practical FET structure was described and analyzed in detail by Shockley in 1952, a year before its reduction to practice by Foy and Wiegmann, working under the direction of Ross and Dacey. This was a case of having metaphorical lightning strike twice in the same place, because as we noted in Section 2. 1, the same inventor had accomplished the same feat just a few years earlier with the BJT! Shockley's FET employed a pair of PN junctions for defining the channel, and hence has become known as the junction FET, or JFET. His innovation placed the region of crucial properties well inside the semiconductor crystal and away from the troublesome and poorly understood surfaces. Varying degrees of reverse bias on these channel-defining junctions caused varying depletion-layer encroachment upon the channel, thus achieving the aim of altering the areal density (number per unit area) of carriers in the thin channel.

While the laboratory JFETs reported in 1953 constituted a feasibility demonstration, they were not practical devices. The JFET requires a degree of control over areal impurity density in its channel that is about ten times better than that required, for example, in the base region of a BJT. Thus the JFET was a device awaiting new and more refined fabrication options. The first practical JFETs, made in 1960, incorporated channels produced by the exponential growth of silicon from a vapor. Today the technology of choice is usually ion implantation, which achieves unprecedented control of areal impurity density in thin layers.

In the discrete-device arena, JFETs have played a relatively small part. Unique properties, especially low noise, have earned for them a "niche market." At first JFETs did not play a role in integrated circuits either, because their peculiar requirements found that a kind of hybrid JFET-MOSFET device could play an advantageous part in circuits dominated by the later. Today, JFETs are being incorporated into even bipolar integrated circuits, thanks to the almost commonplace availability of ion-implantation technology. In addition, the JFET may find application in monocrystalline three-dimensional integrated circuits.

The MOSFET was the second important field-effect device to come along. The heart of its original embodiment was a metal-oxide-silicon sandwich, with initial letters responsible for the acronym MOS, as in MOSFET and MOS technology. The sandwich was in fact a parallel-plate capacitor, with one metal plate and one silicon plate. Thus the structure indeed evokes early (and unsuccessful) experiments that endeavored to modulate the majority-carrier population in a thin semiconductor layer as a way of realizing a solid-state amplifier. But important new features were present in the MOSFET, and these spelled success. First, the new device exploited the infant oxide-growth technology that started with a polished single-crystal silicon sample. Significantly, one of the MOSFET inventors was also a pioneering developer of the oxide-growth technology and student of oxide-silicon interfaces. Second, the MOSFET exploited the creation of an inversion layer in the silicon at the interface, rather than attempting to achieve a significant modulation of majority-carrier areal density in a thin sample. More specifically, the metal plate of the capacitor was used to create a potential well at the surface of the silicon, a well capable of retaining the desired carriers-elections in a P-doped sample or holes in an N-doped sample. This constitutes an alternative description of the inversion layer at an insulator-semiconductor interface. As an isolated entity, this inversion layer is identical to that in a grossly asymmetric junction at equilibrium. But the two cases are different in context. In the present case the carriers are confined within the potential well because they cannot penetrate the insulator on the other side of the interface, while in the PN-junction case, the inversion-layer carriers are in direct communication with the identical majority carriers on the heavily doped side of the junction. Because the MOSFET has overwhelming practical importance today, and because all field-effect devices involve similar principles, our focus in the balance of this chapter will be on the MOSFET.

The FET chronicle by no means ended with the MOSFET. However, a new family was proposed in 1966 wherein control was exercised through a metal-semiconductor junction, or Schottky junction, and reduction to practice was reported the following year. The channel carriers in this device are confined between the depletion region associated with a Schottky junction and that associated with a PN junction. (Sometimes the PN junction is replaced by a junction between an extrinsic region and a semi-insulating region.) This structure was given the name MESFET, for metal-semiconductor field-effect transistor. It has been realized using a number of semiconductor materials, but GaAs is by far the most common. The MESFET, in fact, is the dominant compound-semiconductor transistor today.

A still more recent FET is the MODFET, which designates the modulation-doped FET. It creates a potential well for electrons in a region of extremely light doping, and as a result, the ionic scattering of drifting electrons is reduced to near zero, leading to extreme values of electron mobility, well above 10^4 $cm^2/(V \cdot s)$ at room temperature. (Hole mobilities are typically small in the GaAs family of materials.) By cooling the sample to reduce phonon activity, one can achieve electron mobilities exceeding 10^5 $cm^2/(V \cdot s)$. Such enhanced electron mobilities were first observed in 1978, and were realized in a field-effect transistor in 1980. This feature of the MODFET has given it a second name, HEMT, the acronym for high-electron-mobility transistor. The significance of high carrier mobility is high-speed operation of the resulting device. Still another name for the MODFET is TEGFET, for two-dimensional-electron-gas FET.

The feat of separating channel electrons in the MODFET from the impurity atoms that contribute them is accomplished by means of a hetero-junction, which is an interface between two semiconductor materials of differing energy gap. Two materials often used to form a hetero-junction are the binary compound GaAs and the ternary compound AlGaAs. Hetero-junctions are growing in importance in semiconductor technology, making possible proliferating device innovations. One of these is yet another FET, the HIGFET, which stands for hetero-junction insulated-gate field-effect transistor. Among its advantages is the ready achievement of complementary, or N-channel and P-channel-devices, which make possible superior performance.

A still further advance in the FET art uses hetero-junctions to create a series of quantum wells or a superlattice, terms that describe a set of very thin, parallel layers of differing energy gap in a semiconductor crystal, usually involving binary and ternary compound semiconductors. Layer thickness must be appreciably less than the Broglie wavelength, and typically amounts to a few tens of angstroms. (A superlative can also be created by doping variations, and even by light or sound waves!)

One kind of FET-like structure exploiting quantum-well layers uses them as multiple channels that are under the control of a gate electrode. Because carriers are able to move in the planes of such layers at great velocity, the resulting device is capable of high gain and fast switching. An even more recent proposal would replace the two-dimensional quantum well layers by one-dimensional quantum-well "wires". The transition from classical to quantum principles of operation opens new realms of device possibilities. A substantial number of variations on these and related themes already exist, and it is safe to predict that the future holds many more.

2.2.3 MOSFET Definitions

The essential structure of a MOSFET is shown in Figure 2-8; for purposes of explanation, one can imagine that this structure had been cut out of an integrated circuit. The capacitor like arrangement formed on and by the silicon substrate accounts for the acronym MOS, as was noted above, and as is also noted in the diagram. The metal top plate was made of aluminum in most early devices. But since about 1970, heavily doped polycrystalline silicon (polysilicon) has been favored. Many different insulating materials have been employed through the years, but thermally grown oxide formed in situ is currently and historically by far the most important option. This description of the silicon oxide means that the necessary silicon was supplied by the single-crystal substrate, and its growth was induced, as was noted before, by heating the sample in an oxidizing atmosphere.

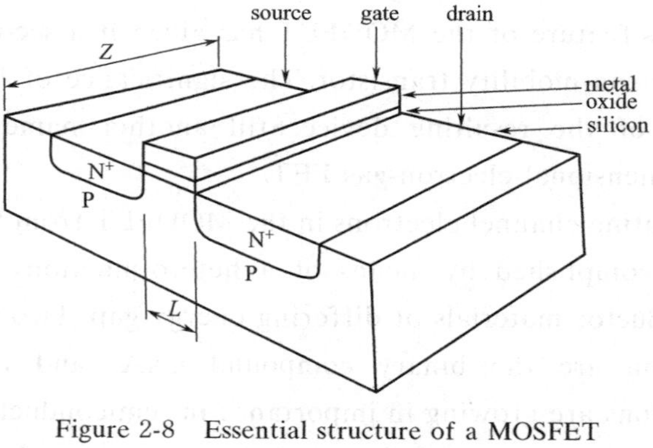

Figure 2-8 Essential structure of a MOSFET

The metal top plate, or field plate, becomes the control electrode of the MOSFET and is known as the gate. Applying a sufficient voltage to the gate with respect to the substrate causes electrons (normally in the minority in the P-type substrate) to be attracted to the interface, thus creating an inversion layer. This thin layer constitutes the resistor cited earlier, and is termed the MOSFET channel. Ohmic contact is made to the ends of the channel by means of heavily doped N-type regions, usually formed by ion implantation or solid-phase diffusion, or some combination of the two.

It is evident in Figure 2-8 that the MOSFET exhibits bilateral symmetry. That is, channel carriers can be made to flow in either direction, depending on bias polarity. This symmetry is important in certain MOSFET applications. The two channel terminals have been named according to which one supply carriers to the channel and which one receives carriers from it. These two terms are source and drain, respectively, and it is evident that bias polarity determines which is which. This choice of terms has been carried over from the JFET, and was made there to provide distinctive initial letters for subscript purposes. (The physicist's classically favored terms source and sink are obviously wanting with respect to this requirement.) To summarize, for the N-channel device under consideration, the more positive of the two channel terminals is the drain, the other is the source.

It is customary to refer all voltages in the MOSFET to the source voltage (just as in an earlier era, terminal voltages in a vacuum tube were referred to that of the cathode). Thus, two important bias values are the drain-source voltage V_{DS} and the gate-source voltage V_{GS}, continuing to use the double-subscript notation. But the substrate constitutes a fourth terminal. And a bias from substrate to source indeed alters device properties; this is a situation that sometimes cannot be avoided, especially in integrated circuits. To deal with it, we need at the most simplistic level a distinctive subscript for "substrate" too, and so the homely term bulk has been adopted as an approximate synonym. It is intended to connote the interior of the substrate. We shall examine the effect of a bulk source voltage V_{BS}. But for the initial description, let us assume that the source and substrate are electrically common, to serve as voltage reference. Further, for the moment let $V_{DS}=0$.

There is a certain voltage V_{GS} at which the hole population immediately under the oxide-silicon interface will be identical to that deep within the bulk of the crystal. Under such conditions, a band diagram representing conditions from the interface to the interior of the P-type silicon crystal would have perfectly horizontal band edges. This magic gate-bias value is termed the flat-band voltage, $V_{GS} = V_{FB}$, and is valuable as a conceptual starting point. When the gate terminal is now made more positive with respect to source (and substrate), the band edges begin to bend near the oxide-silicon interface. This matter will be examined in detail. At the moment, what is important is that potential Ψ_S at the surface of the silicon crystal (at the oxide-silicon interface) is becoming more positive through the influence of the field plate. Hole density at the interface declines and electron density rises. This is dictated by the law of mass action, $pn = n_i^2$, because the silicon remains (for engineering purposes) at equilibrium. The reason for the caveat is that any leakage current through the oxide is accompanied by a net current in the silicon that violates equilibrium. But the superb insulating qualities of SiO_2-resistivity in excess of $10^{15}\ \Omega \cdot cm$ make this a vanishing concern at present.

There exists another critical voltage at which the electron density right at the interface is equal to the whole density in the bulk. This condition is termed the threshold of strong inversion and is specified by the statement $V_{GS} = V_T$. The chosen term is apt, because channel conductance increases steeply with gate voltage for $V_{GS} > V_T$. At the threshold voltage and just below it, the electrons that form the incipient channel contribute a very small amount of drain-source conduction, leading to subthreshold current. For still smaller values of V_{GS}, there exists essentially an open circuit from drain to source because the applied voltage V_{DS} reverse biases the drain junction. Only a tiny fraction of V_{DS} appears as forward bias on the source junction. Thus to a good approximation, one can say that the channel conductance is "turned on" by V_{GS} at the threshold voltage V_T.

2.2.4 Universal Transfer Characteristics

Treating the MOSFET simplistically as a parallel-plate capacitor with a lateral current and a resulting IR drop in one-plate neglects important considerations, as we shall see. This treatment has been accorded to two MOSFETs fabricated simultaneously, but having very different aspect ratios Z/L. Both curves have extensive linear regions. The curvature near the bottom, most evident in the higher-current device, exists because some conduction begins below the threshold voltage V_T, the phenomenon labeled subthreshold conduction. Nonetheless, because the two devices are identical in all respects except aspect ratio, extrapolating the linear portions of their I-V curves leads to a consistent and meaningful theoretical value of V_T. The value of seven volts observed for these early devices is extremely high by today's standards, and was the result of fabrication problems. A typical value today is of the order of one volt.

The device described up to this point is known as the enhancement-mode, or E-mode, MOSFET. It is designated because it has no channel at equilibrium; a gate voltage is required to create the inversion layer that serves as channel, and further increment of gate voltage enhance the conductivity of the channel.

It is possible, however, to create a channel that exists even in the device at equilibrium. This is usually done today by conventional doping of the region just under the gate oxide, with ion implantation being an especially favored technique. The result is a structure somewhat like that shown in cross section in Figure 2-9. The channel is bounded on the bottom by a PN junction, just as in a JFET. This time a negative voltage is applied to the gate, so that the MOS capacitor requires positive charge in the silicon, donor-ion charge fulfills that requirement. In other words, a depletion layer is formed in the N-type channel region. Once again a positive V_{DS} gives rise to an IR drop in the channel. This time it results in a thicker depletion layer at the drain end of the channel than at the source end. For the same reason, the depletion layer of the PN junction below the channel is also thicker at the drain end of the channel. Because the application of a negative voltage to the gate causes current reduction through channel depletion, this kind of device is termed a depletion-mode or D-mode MOSFET. Realized in the way just described, the D-mode MOSFET is a kind of hybrid device, with a MOSFET-like upper boundary for the channel, and a JFET-like lower boundary.

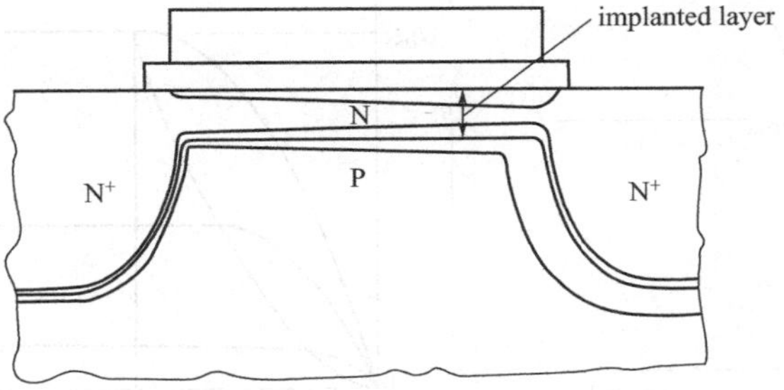

Figure 2-9 Schematic cross-sectional representation of a D-mode N-channel MOSFET

Although the channel region in the D-mode MOSFET is a bit thicker than the inversion-layer channel of the E-mode device, the D-mode MOSFET channel still qualifies as "nearly two-dimensional", and as a result, the D-mode MOSFET is also very nearly a square-law device. Its essentially parabolic current-to-voltage transfer characteristic looks like that for the E-mode device, but is translated leftward along the V_{GS} axis, as can be seen in Figure 2-10 (a). Thus the D-mode device has two structure-determined electrical constants: There is the characteristic current I_{DSS} at which the output current saturates when $V_{GS} = 0$. Then there is the value $V_{GS} = V_P$ that is required to "pinch off" the channel completely, so that no current can flow. Under these conditions, the depletion layer produced by gate voltage has moved down to touch that of the PN junction, all the way from source to drain; as a result, the channel is fully depleted of carriers. The output characteristics that accompany the transfer characteristic in Figure 2-10 (a) are shown in Figure 2-10 (b). Only the parameter labeling of the curves distinguishes them from the E-mode characteristics.

There is an important respect in which the D-mode MOSFET is distinguished from the JFET, with the JFET being also a D-mode device. The D-mode MOSFET possesses an enhancement regime of operation that augments its depletion regime. When positive gate bias V_{GS} is applied, additional electrons are brought into the channel from the source, and channel conductivity increases. This fact is represented in Figure 2-10 (a) by the extension of the transfer characteristic to the right of the I_D axis, or into the regime of positive V_{GS}. The square-law behavior continues. Thus there are additional curves at higher current that can be drawn in Figure 2-10 (b). In the JFET, by contrast, where the channel is bounded both top and bottom by PN junctions, positive bias on the P-type gates (the top and bottom gates are usually common) will forward bias the gate junctions. Thus, for a gate bias of more than a few tenths of a volt, the gate junctions will conduct, and the high input resistance of the device is lost, normally an undesirable situation. As a practical matter, therefore, the transfer characteristic of the N-channel JFET can be extended only a short way into the positive-V_{GS} regime that defines enhancement-mode operation.

Thus far we have considered only N-channel MOSFETs. But it is evident that by reversing all conductivity types, currents, and voltages, we can have E-mode and D-mode P-channel devices. In fact, for technological reasons P-channel devices were dominant early

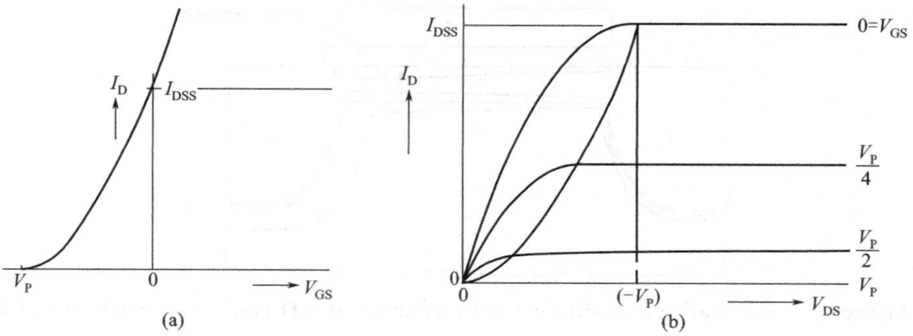

Figure 2-10 Current-voltage characteristics of a D-mode MOSFET

in the MOSFET era. It is possible to construct a transfer-characteristic diagram that summarizes all four possibilities-E and N and P. Normalization once more is convenient. The D-mode device is best treated first because it possesses a characteristic current I_{DSS} as well as a characteristic voltage V_P. Using these quantities for normalization leads to the second-quadrant curve for the D-mode N-channel MOSFET. It is simply a normalized version of the curve shown in Figure 2-10 (a). Since the D-mode P-channel MOSFET involves reversing currents and voltages, we merely construct a second curve in the fourth quadrant, one that is symmetric in the origin to the second-quadrant curve.

While the E-mode device does not have an explicit characteristic current for normalization, it does have an implicit characteristic current $I_D = KV_T^2$. Using this and the voltage $V_{GS} = V_T$ for normalization, we complete the figure, displaying four curves that represent all possible MOSFETs.

Performance advantages result from combining E-mode and D-mode devices in a circuit, and such E-D products are common today. However, the combination of complementary E-mode devices, or N-channel and P-channel E-mode devices, is even more advantageous. In fact, complementary MOSFET circuits, or CMOS circuits, constitute today's most rapidly growing technology.

A number of MOSFET symbols are in common use. Three informative kinds are shown in Figure 2-11. Distinction between a D-mode and E-mode device is sometimes maintained by using a solid line to represent the channel in the former case, and a broken line as in Figure 2-11 (a) for the E-mode case. Another symbol, as in Figure 2-11 (b), acknowledges that the substrate constitutes a fourth terminal of the MOSFET. Consistent with conditions in this section, we show the substrate connected (with a dotted line that is not part of the symbol) to the source. This symbol also indicates an N-channel device, with the arrow direction, as usual, from P to N, or from the P-type substrate to the channel. Finally, one sometimes wishes to denote a "preferred source" in spite of the intrinsic symmetry of the MOSFET, and this can be done by offsetting the gate terminal in the source direction.

A third symbol in common use is show in Figure 2-11 (c). Its virtues are simplicity and parallelism with the BJT symbol. That is, an arrow is used to designate the source terminal and also to indicate normal current direction, just as an arrow designates the BJT emitter

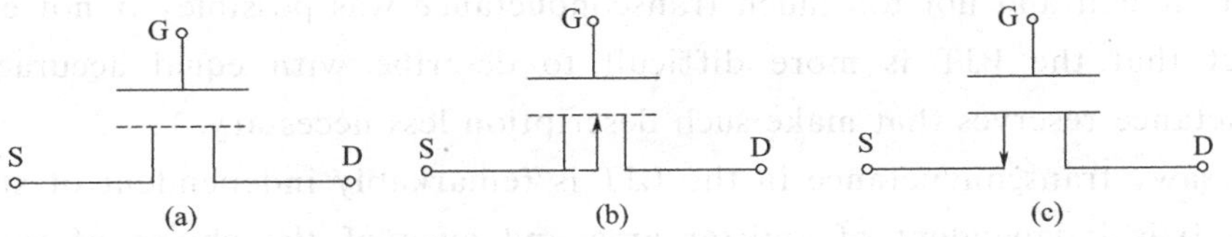

Figure 2-11 MOSFET symbols

terminal and its normal current direction. The line connecting source and drain of course peppermints the channel, and is sometimes made thicker to distinguish a D-mode form and E-mode device.

2.2.5 Transconductance

The rate of change of output current with respect to input voltage is an important characteristic property of a device and is termed transconductance. Its symbol is g_m. For the MOSFET

$$g_m = \frac{\partial I_D}{\partial V_{GS}} \tag{2.10}$$

The partial derivative has conductance dimensions and has "across" or trans properties, to repeat, because it describes a result at the output port produced by a change at the input port. Transconductance exhibited in the saturation regime is of particular importance, and is of course simpler than for the curved regime.

$$g_m = 2K(V_{GS} - V_T) \tag{2.11}$$

Experimental data confirms this simple linear relationship, once again drawing data from P-channel devices of the early 1960s.

Transconductance has great importance because it plays a large part in determining the switching speed of an active device. High-transconductance devices yield circuits capable of high-speed operation. Eq. (2.11) makes it evident that MOSFET transconductance is fixed by structure and technology. That is, it depends on properties more or less under the designer's control. This explains the preference for "N-MOS" over "P-MOS" to take advantage of the higher electron mobility. It is partly responsible for the constantly declining values of X_{ox} in recent decades (thus increasing C_{ox}). Finally, the presence of the factor Z/L in g_m means that the designer makes a g_m adjustment at the graphics console (or in an earlier era, at the drawing board). The designer's aim, however, is to make g_m just large enough and no larger, because overdesign would involve an oversized MOSFET, requiring more silicon area. Circuit costs are directly tied to circuit area.

This brings us to another fact that helped the MOSFET toward its present dominance. Equations of simple form, and coming from the simplistic analysis just presented, give an excellent description of MOSFET properties after empirical adjustment. As a result, computer design and optimization of MOS integrated circuits has been practical since the mid-1960s. Hence the necessary fine-tuning of device sizes needed to give just enough and not too much transconductance was possible, if not easy. (It is a curious fact that the BJT is more difficult to describe with equal accuracy, but has transconductance reserves that make such description less necessary.)

As we saw, transconductance in the BJT is remarkably independent of structure and technology. It is independent of emitter area and even of the choice of semiconductor material! This is because BJT operation involves modulation of a naturally occurring potential barrier with invariant incremental properties. But in the MOSFET case, one is designing a variable resistor. Many BJT-MOSFFET transconductance comparisons have been made over the years. These are not straightforward, because the two devices are so different, a subject addressed more fully in section, but they typically indicate that the BJT is superior in transconductance by a factor ranging from 10 to 100. The fact that the MOSFET is today's dominant device is the result of a large collection of other MOSFET attributes and advantages that will be treated later. The designer, seeking always to have the best of both worlds, is today introducing the BJT into MOSFET circuits-especially CMOS circuits-in places where high transconductance is crucial, and the resulting BiCMOS products are receiving increasing attention.

2.2.6 SPICE Models

There are three primary SPICE MOSFET models, identified as Level 1, 2, and 3. Level 1 employs the rudimentary analysis of Section 2.2.1, with empirical adjustment. The most important correction is a coefficient applied to the current expressions to bring saturation-regime current into congruence with experiment and more accurate analyses. The result of such adjustment is shown in figure, where the rudimentary model is brought into good agreement with the ionic-charge model.

The Ihantola-Moll, or ionic-charge, analysis is taken as the Level 2 SPICE mode. It is used in a form general, permitting nonzero values of bulk-source voltage. A notable departure of the Level 2 from the Level 1 analysis is a three-halves-power dependence of current on voltage, rather than the square-law dependence in the Level 1 case. Because the numerical handling of a fractional power is more time-consuming than that of an integral power, circuit designers were motivated to seek empirical modifications of the Level 2 model. In the resulting SPICE Level 3 model, they not only removed this shortcoming but also were able to deal empirically with many other effects as well, such as short-channel and channel-shortening effects. The latter is attributable to the variations of drain-junction depletion-layer thickness, and is analogous to the Early effect in the BJT. The former implies accounting for some 3D complications, these inevitably arise when channel length approaches depletion-layer dimensions.

The semiempirical Level 1 and Level 3 models are especially advantageous for circuit analysis. But because our present interest is equally on device physics and circuit analysis, we will focus primarily on the Level 2 Ihantola-Moll model. The starting point is a generalized form for equation that takes into account the effect of finite bulk-source voltage

$$I_D(V_{GS}, V_{BS}, V_{DS}) = 2K\{[V_{GS} - V_{FB}^* + 2\Psi_B - (V_{DS}/2) V_{DS} + 2\gamma/3] \\ (V_{DS} - 2\Psi_B - V_{BS})^{3/2} - (-2\Psi_B - V_{BS})^{3/2}\} \quad (2.12)$$

where

$$K = (1/2)\mu_n C_{ox} Z/L^* \quad (2.13)$$

and L^* is an elaborated version of channel length L that takes into account a technological channel-shortening effect. Impurities implanted to form the source and drain regions diffuse laterally during subsequent heat treatments by the amount L_D^*, and as a result, the channel is shortened by the amount $2L_D^*$. Hence

$$L^* = L - 2L_D^* \quad (2.14)$$

Our purpose here is to introduce the large number of possible input parameters for the Level 2 SPICE model. To compare the model with experiment, we will employ the device geometry and the current-voltage characteristics shown in Figure 2-12. Also given in the latter figure are the characteristics calculated using the Ihantola-Moll model. Figure 2-13 (a) presents a static equivalent-circuit model for the ideal MOSFET described by Eq. (2.12). The current generator ID is a function of the three variables V_{GS}, V_{BS}, and V_{DS}. Figure 2-13 (b) adds to the equivalent circuit the parasitic elements present in a real MOSFET, needed for satisfactory dynamic analysis of the device.

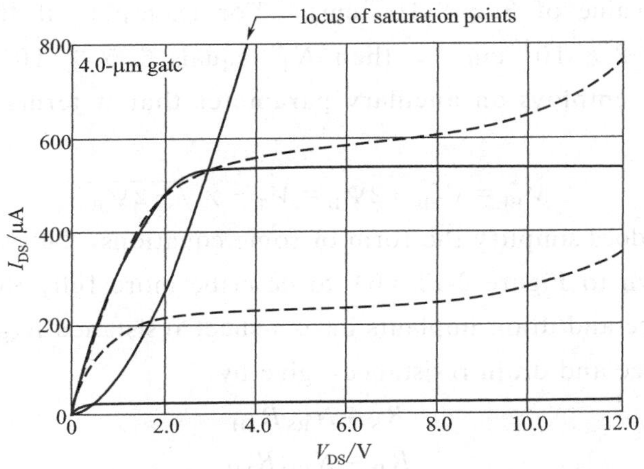

Figure 2-12 Three output characteristics for MOSFET

The oxide thickness X_{ox} determines the capacitance per unit area C_{ox}, and along with the surface mobility μ_n, enters into the current coefficient K. The substrate doping N_A and the specific capacitance C_{ox} enter the body-effect parameter that we repeat here

$$\gamma = \sqrt{(2\in qN_A/C_{ox})} \quad (2.15)$$

The substrate doping N_A also determines the bulk potential W_a, since

$$\Psi_B = (K_t/q)\ In(N_A/n_i) \quad (2.16)$$

The mobility parameters (n is used to define an effective mobility) are described later. In addition, the channel-length modulation factor, λ, is similar to the base-thickness modulation parameter in the BJT model. The effective oxide fixed change density N_f

empirically fixes the zero-order threshold voltage determined by HD, the barrier-height difference. For metal field plates, the SPICE values agree exactly with the experimental values given by Deal and Snow, so that

$$H_D = \pm \Psi_B + 0.61\text{V} \tag{2.17}$$

But for polysilicon field plates, we will use "effective" values N_f^* and H_D^* because the SPICE values differ slightly from the experimental results of Weber. The approximate SPICE values were chosen for convenient symmetry, with the barrier-high difference stated as

$$H_D^* = \pm [\Psi_B - (0.5575\text{V})\ T_G] \tag{2.18}$$

The positive sign represents a P-type substrate, and the negative sign, an N-type substrate. The type-of-gate parameter T_G is positive for a polysilicon gate when the gate-doping type is opposite to that of the substrate, and negative, when the two types are the same. The effective flat-band voltage V_{FB}^* is given by

$$V_{FB}^* = H_D^* - (qN_f^* / C_{ox}) \tag{2.19}$$

And finally the threshold voltage is given by

$$V_T = V_{FB}^* - 2\Psi_B + \gamma\sqrt{-2\Psi_B} \tag{2.20}$$

To compensate for the departure of the SPICE Eq. (2.18) from the more exact, we add a constant value of $3.3 \times 10^{10}\text{cm}^{-2}$. For example, if the actual fixed-oxide charge N_f equals $3.2 \times 10^{10}\text{cm}^{-2}$, then N_f^* equals $6.5 \times 10^{10}\text{cm}^{-2}$. The internal SPICE program also employs an ancillary parameter that it terms the effective barrier height, defined as

$$V_{BI}^* = V_{FB}^* - 2\Psi_B = V_T - \gamma\sqrt{-2\Psi_B} \tag{2.21}$$

This parameter does simplify the form of some equations, but we will not use it.

Now let us return to Figure 2-13 (b) to describe more fully the various elements it introduces. The source and drain implants have a sheet resistance R_{SH} that causes parasitic series source resistance and drain resistance, give by

$$R_S = N_{RS} R_{SH} \tag{2.22}$$

and

$$R_D = N_{RD} R_{SH} \tag{2.23}$$

where N_{RS} and N_{RD} are the numbers of resistive squares associated with each implanted region. The source and drain implants also produce PN junctions with a saturation-current density (extrapolated to zero bias) of J_O. The saturation currents for the corresponding junctions are given by

$$I_{BS} = A_S J_O \tag{2.24}$$

and

$$I_{BD} = A_D J_O \tag{2.25}$$

where A_S and A_D are areas of the source and drain implants. In normal operation, the source and drain junctions are reverse-biased so that these diodes contribute only relatively unimportant leakage currents. However, the associated depletion-layer capacitances are important, and are divided into sidewall portions and bottom portions. The zero-bias values of the bottom parts are given by

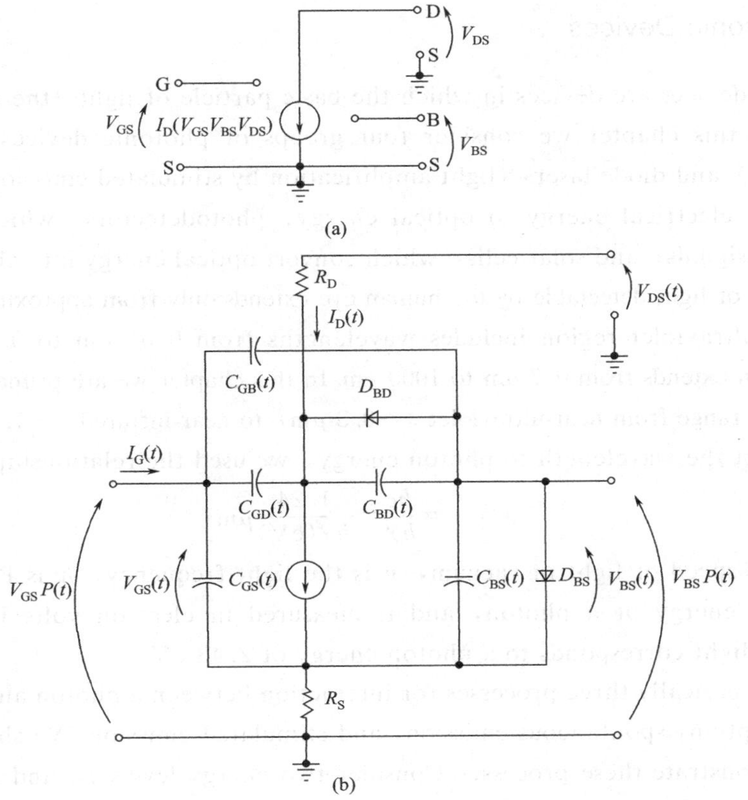

Figure 2-13 Level 2 modeling of the MOSFET

$$C_{OBS.\,B} = A_S C_{0.\,B} \tag{2. 26}$$

and

$$C_{OBS.\,B} = A_D C_{0.\,B} \tag{2. 27}$$

where $C_{0.\,B}$ is the junction capacitance per unit area. The sidewall portions of the drain and source junctions contribute capacitances given by

$$C_{OBS.\,SW} = P_S C_{0.\,SW} \tag{2. 28}$$

and

$$C_{OBD.\,SW} = P_D C_{0.\,SW} \tag{2. 29}$$

where $C_{0.\,SW}$ is the sidewall capacitance per unit length, P_S is the length of the sidewall for the source region, and P_D is the corresponding length for the drain region.

2. 3 Microwave and Photonic Devices

2. 3. 1 Microwave Devices

The microwave frequencies cover the range from about 1 GHz (10^9 Hz) to 1000 GHz, with corresponding wavelengths from 30 to 0. 03 cm. The frequencies from 30 to 300 GHz are called the millimeter wave band, because the wavelength is between 10 and 1 mm; higher frequencies are called the submillimeter wave band.

2.3.2 Photonic Devices

Photonic devices are devices in which the basic particle of light—the photon—plays a major role. In this chapter we consider four groups of photonic devices: light-emitting diodes (LEDs) and diode lasers (light amplification by stimulated emission of radiation), which convert electrical energy to optical energy; photodetectors, which electronically detect optical signals; and solar cells, which convert optical energy into electrical energy.

The range of light detectable by the human eye extends only from approximately 0.4 μm to 0.7 μm. The ultraviolet region includes wavelengths from 0.01 μm to 0.4 μm, and the infrared region extends from 0.7 μm to 1000 μm. In this chapter we are primarily interested in the wavelength range from near-ultraviolet (~0.3 μm) to near-infrared (~1.5 μm).

To convert the wavelength to photon energy, we used the relationship

$$\lambda = \frac{c}{v} = \frac{hc}{h\gamma} = \frac{1.24}{h\gamma(\mathrm{eV})}\mu\mathrm{m} \tag{2.30}$$

where c is the speed of light in vacuum, v is the light frequency, h is Planck constant, and $h\gamma$ is the energy of a photon, and is measured in electron volts. For example, a 0.5 μm green light corresponds to a photon energy of 2.48 eV.

There are basically three processes for interaction between a photon and an electron in a solid: absorption, spontaneous emission, and stimulated emission. We shall use a simple system to demonstrate these processes. Consider two energy levels E_1 and E_2 of an atom, where E_1 corresponds to the ground state and E_2 corresponds to an excited state (Figure 2-14), any transition between these states involves the emission or absorption of a photon with frequency $h\gamma_{12}$ given by $h\gamma_{12} = E_2 - E_1$. At room temperature, most of the atoms in a solid are at the ground state. This situation is disturbed when a photon of energy exactly equal to $h\gamma_{12}$ impinges on the system. An atom in state E_1 absorbs the photon and thereby goes to the excited state E_2. The change in the energy state is the absorption process, shown in Figure 2-14 (a). The excited state of the atom is unstable; and after a short time, without any external stimulus, it makes a transition to the ground state, giving off a photon of energy $h\gamma_{12}$. This process is called spontaneous emission [Figure 2-14 (b)]. When a photon of energy $h\gamma_{12}$ impinges on an atom while it is in the excited state [Figure 2-14 (c)], the atom can be stimulated to make a transition to the ground state and gives off a photon of energy $h_{\mathrm{V}12}$, which is in phase with the incident radiation. This process is called stimulated emission. The radiation from stimulated emission is monochromatic because each photon has an energy of precisely $h_{\mathrm{V}12}$ and is coherent because all photons emitted are in phase.

The dominant operation process for the light-emitting diode (LED) is spontaneous emission; for the laser, it is stimulated emission; and for the photo-detector and the solar cell, it is absorption.

Let us assume that the instantaneous populations of E_1 and E_2 are n_1 and n_2, respectively. Under a thermal-equilibrium condition and for $(E_2 - E_1) > 3kT$, the

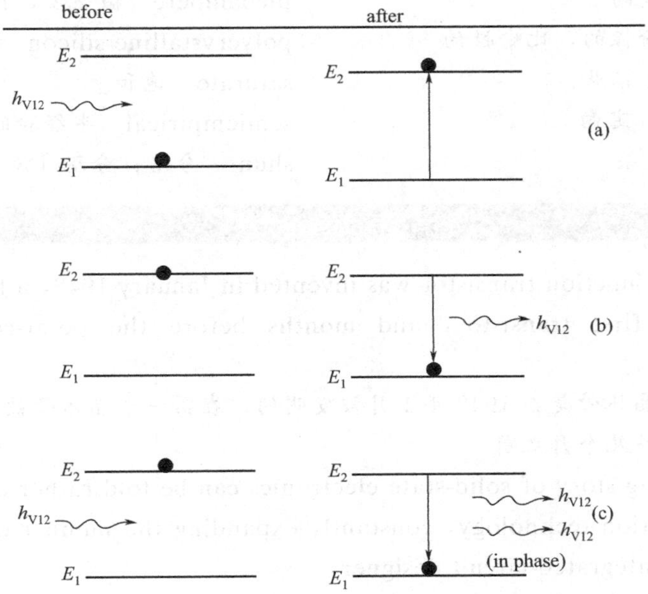

Figure 2-14 The three basic transition processes between two energy levels

population is given by the Boltzmann distribution

$$\frac{n_2}{n_1} = e^{-(E_2 - E_1)/kT} = e^{-h\gamma_{12}/kT} \tag{2.31}$$

The negative exponent indicates that n_2 is less than n_1 in thermal equilibrium; that is, most electrons are at the lower energy level.

In steady state, the stimulated-emission rate (i. e., the number of stimulated-emission transitions per unit time) and the spontaneous-emission rate must be balanced by the rate of absorption to maintain the population's n_1 and n_2 constant. The stimulated-emission rate is proportional to the photon field energy density ρ ($h\gamma_{12}$), which is the total energy in the radiation field per unit volume per unit frequency. Thus, the stimulated-emission rate can be written as $B_{21} n_2 \rho$ ($h\gamma_{12}$), where n_2 is the number of electrons in the upper level and B_{21} is a proportionality constant. The spontaneous-emission rate is proportional only to the population of the upper level and can be written as $A_{21} n_2$ where A_{21} is a constant. The absorption rate is proportional to the electron population at the lower level and to ρ ($h\gamma_{12}$); this rate can be written as $B_{12} n_2 \rho$ ($h\gamma_{12}$), where B_{12} is a proportionality constant. Therefore, we have at steady state.

New Words and Expressions

angstrom 埃
conductance 电导
congruence 相合性；一致
depletion 损耗
Early effects 厄尔利效应
filament 细丝；灯极
hetero-junction 异质结
hybrid 混合电路
incipient 初始的
lumped-element model 集总模型

metallurgical 冶金的

metaphorical 隐喻性的，比喻性的

noise margin 噪声容限

orientation 定位；定向

pedestal 基座；支架

picoampere 微微安，10^{-12} 安培

polycrystalline silicon 多晶硅

saturate 饱和

semiempirical 半经验的

shunt 分路；分路［流］器；并联

Translation

1. The bipolar junction transistor was invented in January 1948, a few weeks after the invention of the first transistor, and months before the point-contact device was announced.

翻译： 双极结晶体管是在1948年1月被发明的，在第一个晶体管被发明几个星期之后，在公布点接触式器件几个月之前。

2. The unfolding story of solid-state electronics can be told rather completely in terms of evolving fabrication technology, constantly expanding the number of options available to the device and integrated-circuit designer.

翻译： 固体电子学发展的故事可以从不断发展的制造技术、不断增多的器件选择和不断增大的集成电路设计者数量方面完整展开。

3. Once the emitter junction has been so identified and labeled, the polarity of the collector junction becomes evident.

翻译： 一旦发射极结经过识别和标记，集电极结的极性就变得显而易见了。

4. It is evident from the dimensions shown that the active portion of the device is confined to a small portion of the overall crystal volume.

翻译： 从平面图可以明显看出，器件的活性部分仅限于整个晶体体积的一小部分。

5. Operation of the BJT can be approached by examining the properties of its two junctions individually, considering them to be isolated.

翻译： 双极结型晶体管的活动可以通过单独检查其两个结的特性来进行，这两个结可以认为是孤立存在的。

6. Then we shall combine the junctions in order to examine their interaction.

翻译： 然后我们将结合这些连接检查它们之间的相互作用。

7. For a combination of reasons, calculations based upon idealized structures can provide meaningful descriptions of device properties and worthwhile guides for design modifications.

翻译： 由于各种原因，基于理想化结构的计算可以为器件特性提供有意义的描述，并为设计修改提供有价值的指导。

8. It is a curious fact, once more, that modern technology is moving ever closer to the step-junction ideal employed in the simplest analyses, with uniform doping in the emitter and collector regions, and sometimes in the base region as well.

翻译： 奇怪的是，现代技术越来越接近阶梯式的转折点，在最简单的分析中采用理想的方法，在发电极和集电极中均匀掺杂的区域，有时也包括底边区域。

9. The relative magnitudes of the secondary currents are represented qualitatively by

their widths, but for clarity all of these widths are grossly exaggerated with respect to that of the primary arrow.

翻译： 二次电流的相对大小用它们的宽度来定性地表示，但是为了清楚起见，所有这些宽度相对于主电流的宽度夸大很多。

10. In spite of the relatively small size of the base current, it is of extreme importance because, as we shall see, it is the control current that is supplied to the input port in the common-emitter BJT.

翻译： 尽管基极电流相对较小，但它是极其重要的，因为它是供给公共发射极 BJT 输入端口的控制电流。

11. It provides one of the sharpest contrasts between the modern silicon transistor and the early germanium transistor.

翻译： 它是现代硅晶体管和早期锗晶体管之间最鲜明的对比之一。

12. Their approach to this challenge employed a variation and extension of charge-control techniques.

翻译： 他们采用了变化和扩展的电荷控制技术来解决这一问题。

Unit 3
Holography and Fourier Optics

3.1 Holography

Holography or wavefront reconstruction was invented before 1948, about twenty years before the development of the laser. Its real success came only after the existence of highly coherent sources suggested the possibility of separating the reference and object beams to allow high-quality reconstructions of any object.

3.1.1 Principle of Holography

In Gabor's original holography, light was filtered and passed through a pinhole to bring about the necessary coherence. The source illuminated a small, semi-transparent object O that allowed most of the light to fall undisturbed on a photographic plate H. In addition, light scattered or diffracted by the object also falls on the plate, where it interferes with the direct beam or coherent background. The resulting interference pattern may be recorded on the plate and contains enough information to provide a complete reconstruction of the object.

To find the intensity at H we may write the field arriving at H as

$$E = E_i + E_0 \tag{3.1}$$

where E_i is the field due to the coherent background and E_0 is the field scattered from the object. The scattered field E_0 falling on H is not simple, both amplitude and phase vary greatly with position. We therefore write

$$E_0 = A_0 e^{i\Psi_0} \tag{3.2}$$

where E_i and Ψ_0 are implicitly functions of position. We write a similar expression for E_i, even though E_i is usually just a spherical wave with nearly constant amplitude A_i. The field falling on the plate may then be expressed as

$$E = e^{i\Psi_i}[A_i + A_0 e^{i(\Psi_0 - \Psi_i)}] \tag{3.3}$$

and the intensity

$$I = A_i^2 + A_0^2 + A_0 A_i e^{(\Psi_0 - \Psi_i)} + A_0 A_i e^{-i(\Psi_0 - \Psi_i)} \tag{3.4}$$

For our purpose it is convenient to characterize the photographic plate H by a curve of amplitude transmittance t_a vs exposure ζ rather than by the D vs $\lg\zeta$ curve of conventional photography. The curve is nearly linear over a short region; we call the slope there β.

In that region, the equation for the t_a vs ζ curve can be written as

$$t_a = t_0 - \beta\zeta \tag{3.5}$$

for the (negative) emulsion.

If we take the exposure time to bet, we find the amplitude transmittance of the developed plate to be

$$t_a = t_0 - \beta t[A_i^2 + A_0^2 + A_0 A_i e^{(\Psi_0 - \Psi_i)} + A_0 A_i e^{-i(\Psi_0 - \Psi_i)}] \tag{3.6}$$

The term A_0^2 contributes to noise in the reconstruction, but we drop it here because of our assumption that the scattered field is small compared with the coherent background.

We may now remove the object and illuminate the developed plate with the original reference beam E_i. The developed plate is known as the hologram. The transmitted field E_t just beyond the hologram is

$$E_t = A_i e^{i\Psi_i} t_a \tag{3.7}$$

and the interesting part is

$$-\beta t A_i e^{i\Psi_i}[A_i^2 + A_0 A_i e^{i(\Psi_0 - \Psi_i)} + A_0 A_i e^{-i(\Psi_0 - \Psi_i)}] \tag{3.8}$$

We may extract factor A_t from the square bracket. Apart from real constants, the result is

$$e^{i\Psi_i}[A_i + A_0 e^{i(\Psi_0 - \Psi_i)} + A_0 e^{-i(\Psi_0 - \Psi_i)}] \tag{3.9}$$

The first two terms here are identical with the field that exposed the plate; the first term corresponds to the coherent background and the second to the wave scattered by the object. An observer looking through the plate would therefore seem to see the object located in its original position. Except for its intensity, this reconstruction is theoretically identical to the object.

The third term is identical with the second term, apart from the sign of the phase term $(\Psi_0 - \Psi_i)$. This term corresponds to a second reconstruction, located on the opposite side of the plate. This conjugate reconstruction is always present and is out of focus when we focus on the primary reconstruction. Its presence therefore degrades the primary reconstruction and is a major obstacle to the production of high-quality holograms. For practical purposes, this obstacle was removed when the laser provided sufficient coherence to separate the object wave from the coherent background with a prism or beam splitter.

3.1.2 Classification of Holograms

A hologram recorded on a photographic plate and processed normally is equivalent to a grating with a spatial varying transmittance. However, with suitable processing, it is possible to produce a spatial varying phase shift. In addition, if the thickness of the recording medium is large compared to the fringe spacing, volume effects are important. In an extreme case, it is even possible to produce holograms in which the fringes are planes running almost parallel to the surface of the recording material; such holograms can reconstruct an image in reflected light.

Based on these characteristics, holograms recorded in a thin recording medium can be divided into amplitude holograms and phase holograms. Holograms can be recorded in relatively thick recording media or can be classified either as transmission amplitude holograms, transmission phase holograms, reflection amplitude holograms, or reflection phase holograms.

3.1.3 Rainbow Holography

A multicolor image can be produced by a hologram recorded with three suitably chosen wavelengths. The resulting recording can be considered as made up of three incoherently superposed holograms. When it is illuminated once again with the three wavelengths used to make it, each of these wavelengths is diffracted by the hologram recorded with it to give a reconstructed image in the corresponding color. The superposition of these images yields a multicolor reconstruction.

The widely used lasers for rainbow holography are the He-Ne laser ($\lambda = 633$ nm) and the Ar^+ laser with two strong output lines ($\lambda = 514$ nm and 488 nm). There are also other laser lines can be used which provide a better choice. One of these is the He-Cd laser line ($\lambda = 422$ nm), which is very attractive as a blue primary but involves the use of one more laser. The Kr^+ laser can be used for recording large holograms, because the power available is much higher, and single-mode output can be obtained with an etalon. The use of the Kr^+ laser line at 521 nm, or the output from a frequency-doubled Nd: YAG laser ($\lambda = 532$ nm), as the green primary has been found to give much better yellow images.

(1) A typical method to generate the color rainbow hologram

Figure 3-1 shows the way to generate the color rainbow hologram in a single step. After those three light expose the recording holograms for three times respectively, the rainbow hologram can be formed in the dry plate. When this multiplexed hologram is illuminated with a light source, it reconstructs three superimposed images of the object.

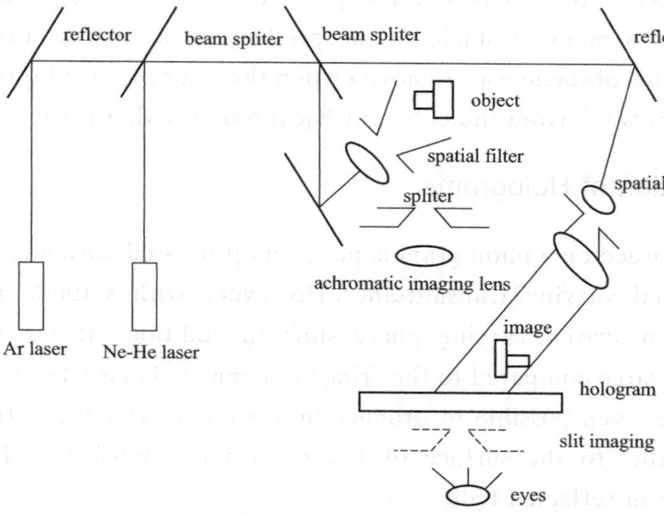

Figure 3-1 Color rainbow holography in a single step

(2) The cross-talk problem

A problem in multicolor holography is that each hologram diffracts, in addition to light of the wavelength used to record it, the other two wavelengths are as well. As a result, a total of nine primary images and nine conjugate images are produced. Three of these give rise to a full-color reconstructed image at the position originally occupied by the object. The remaining images resulting from light of one wavelength diffracted by a component hologram recorded with another wavelength are formed in other positions and overlap with and degrade the multicolored image.

Several methods have been tried to eliminate these cross-talk images, including spatial-frequency multiplexing, spatial multiplexing or coded reference waves, and division of the aperture field. However, all these methods suffer from drawbacks such as a restricted image field, a reduction in resolution, or a decrease in a signal-to-noise ratio (SNR). In addition, they need multiple laser wavelengths (or equivalent monochromatic light sources) to illuminate the hologram.

(3) Volume holograms

The first method to eliminate cross-talk that did not involve such penalties was based on the use of volume holograms. A hologram recorded with several wavelengths in a thick medium contains a set of regularly spaced fringe planes for each wavelength. When this hologram is illuminated once again with the original multi-wavelength reference beam, each wavelength is diffracted with maximum efficiency by the set of fringe planes created originally by it, producing a multicolored image. However, the cross-talk images are severely attenuated since they do not satisfy the Bragg condition.

This principle was first applied to produce a two-color hologram of a transparency, and subsequently extended to three-color imaging of diffusely reflecting objects. The optical setup is shown schematically in Figure 3-2.

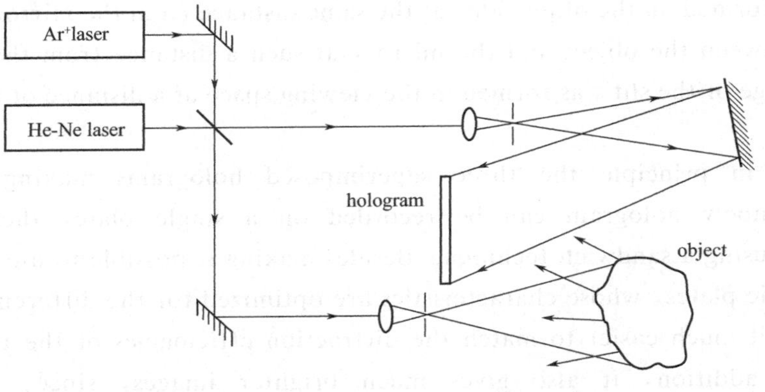

Figure 3-2 Setup used to record a multicolor hologram of a diffusely reflecting object in a thick recording medium

Blue and green light ($\lambda = 488$ nm and 514 nm) from an Ar^+ laser was mixed with red light ($\lambda = 633$ nm) from a He-Ne laser to produce two beams containing light of all three wavelengths. One beam was used to illuminate the object while the other was used as a reference beam, and the resulting hologram was recorded in a thick photographic emulsion. When this hologram was illuminated once again with a similar multicolor beam at the appropriate angle, a multicolor reconstructed image with negligible cross-talk was obtained.

(4) Multicolor rainbow holograms

A completely different approach was opened up by the extension of the rainbow hologram technique to three-color recording. This made it possible to produce holograms that reconstruct very bright multicolor images when illuminated with a white-light source. Multicolor rainbow holograms can be produced in both one single step and two steps. Here a typical optical system for this purpose, using a concave mirror, is shown in Figure 3-3.

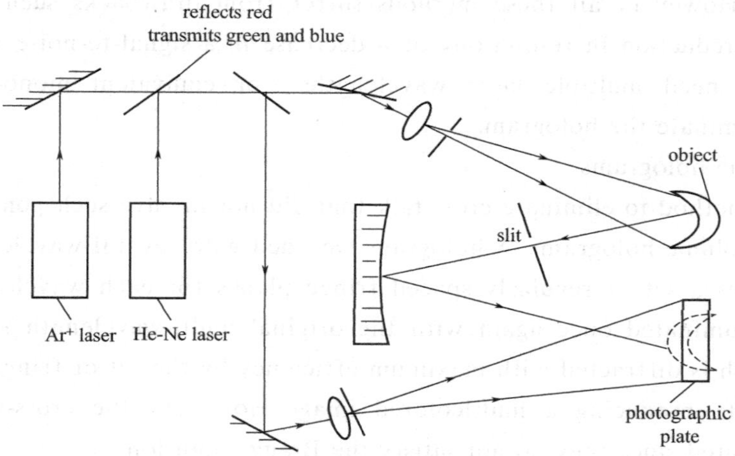

Figure 3-3 Layout of the optical system used to produce multicolor rainbow holograms in a single step with a concave mirror

The object, turned sideways, was placed on one side of the axis of the mirror so that its image was formed on the other side, at the same distance from the mirror. A vertical slit was placed between the object and the mirror, at such a distance from the mirror that a magnified image of the slit was formed in the viewing space at a distance of about 1 m from the hologram.

Although in principle the three superimposed holograms making up the final multicolor rainbow hologram can be recorded on a single plate, there are several advantages in using a sandwich technique. Besides making it possible to use different types of photographic plates, whose characteristics are optimized for the different wavelengths, it also makes it much easier to match the diffraction efficiencies of the three individual holograms. In addition, it also gives much brighter images, since, with bleached holograms, the loss in diffraction efficiency due to multiplexing three holograms on a single plate can be partially avoided.

Multicolor holograms give bright images even when illuminated with an ordinary tungsten lamp. In addition, the images exhibit high color saturation and are free from cross-talk. Problem with emulsion shrinkage are eliminated, since volume effects are not involved. As with any rainbow hologram, the colors of the image change with the viewing angle in the vertical plane. This change can be utilized effectively in some types of displays, but, where necessary, it can be kept within acceptable limits by optimization of the length of the spectra projected into the viewing angles in the vertical plane.

3. 1. 4 Computer-generated Holography

Holograms generated by means of a computer or computer-generated holography (CGH), as shown in Figure 3-4, can be used to produce wavefronts with any prescribed amplitude and phase distribution; they are therefore extremely useful in applications such as laser-beam scanning and optical filtering as well as for testing optical surfaces.

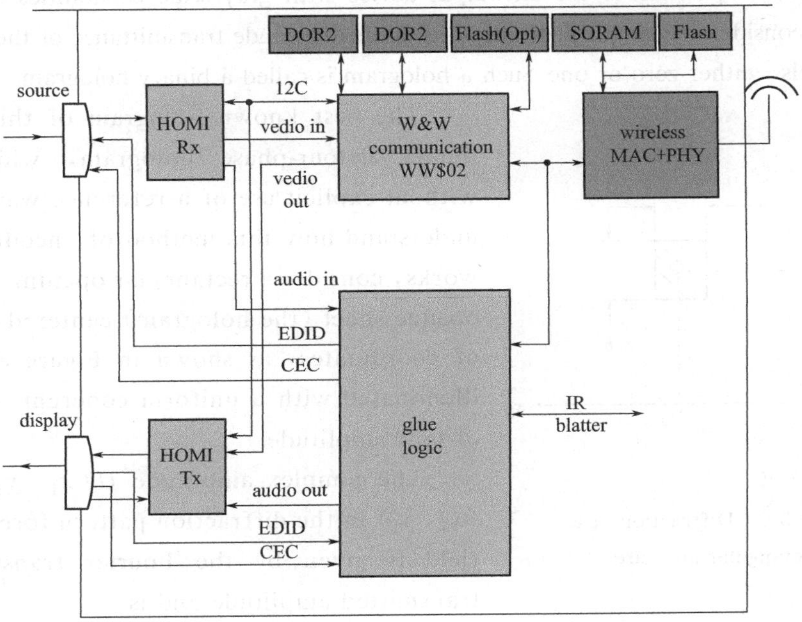

Figure 3-4 The principle of CGH

One of the most popular approaches to generate CGH involves two principal steps. The first step is to calculate the complex amplitude of the object wave at the hologram plane. For convenience, this is usually taken to be the Fourier transform of the complex amplitude in the object plane. It can be shown, by means of the sampling theorem, that if the object wave is sampled at a sufficiently large number of points, this can be done with no loss of information. Thus, if an image consisting of $N \times N$ resolvable elements is to be reconstructed, the object wave is sampled at $N \times N$ equally spaced points, and the $N \times N$ complex coefficients of its discrete Fourier transform are evaluated. This can be done quite easily with a computer program using the fast Fourier transform (FFT) algorithm for arrays containing as many as 1024×1024 points.

The second step involves using the computed values of discrete Fourier transform to produce a transparency (the hologram), which reconstructs the object wave when it is suitably illuminated. An alternative approach, which is possible only with a computer-generated hologram, is to produce a transparency that records both the amplitude and the phase of the object wave in the hologram plane. This can be thought of as the superimposition of two transparencies, one of constant thickness with a transmittance proportional to the amplitude of the object wave, and the other with thickness variations corresponding to the phase of the object wave, but no transmittance variations. Such a hologram has the advantage that it forms a single, on-axis image. In either case, the computer is used to control a plotter that produces a large scale version of the hologram. This master is photographically reduced to produce the required transparency.

(1) Binary detour-phase hologram

Although it is possible to use an output device with gray scale capabilities to produce the hologram, a considerable simplification results if the amplitude transmittance of the hologram has only two levels—either zero or one. Such a hologram is called a binary hologram.

The best known hologram of this type is the binary detour-phase hologram, which is made without explicit use of a reference wave or bias. To understand how this method of encoding the phase works, consider a rectangular opening ($a \times b$) in an opaque sheet (the hologram) centered on the origin of coordinates, as shown in Figure 3-5, which is illuminated with a uniform coherent beam of light of unit amplitude.

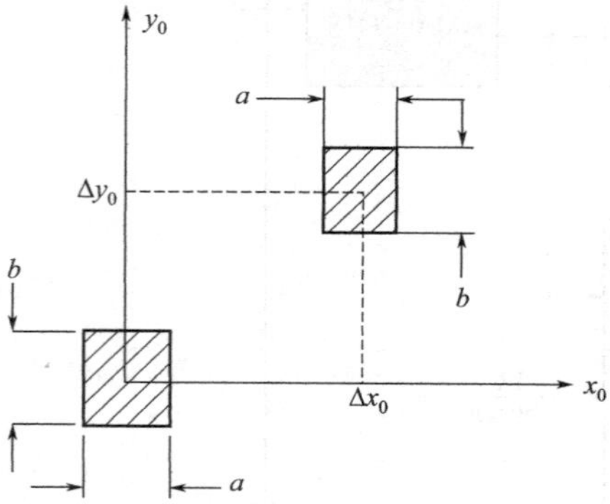

Figure 3-5 Diffraction at a rectangular aperture

The complex amplitude $U(x_i, y_i)$ at a point (x_i, y_i) in the diffraction pattern formed in the far field is given by the Fourier transform of the transmitted amplitude and is

$$U(x_i, y_i) = ab \sin c(by_i/\lambda z) \tag{3.10}$$

We now assume that the center of the rectangular opening is shifted to a point $(\Delta x_0, \Delta y_0)$, and the sheet is illuminated by a plane wave incident at an angle. If the complex amplitude of the incident wave at the sheet is $\exp[i(\alpha \Delta x_0 + \beta \Delta y_0)]$, the complex amplitude in the diffraction pattern becomes

$$\begin{aligned} U(x_i, y_i) &= ab \sin c(ax_i/\lambda z) \sin c(by_i/\lambda z) \exp\left[i\left(\alpha + \frac{2\pi x_i}{\lambda z}\right)\Delta x_0 + i\left(\beta + \frac{2\pi y_i}{\lambda z}\right)\Delta y_0\right] \\ &= ab \sin c(ax_i/\lambda z) \sin c(by_i/\lambda z) \exp[i\alpha\Delta x_0 + i\beta\Delta y_0] \exp\left[i\left(\frac{2\pi x_i}{\lambda z}\right)\Delta x_0 + \frac{2\pi y_i}{\lambda z}\Delta y_0\right] \end{aligned} \tag{3.11}$$

If $ax_i \ll \lambda z$, $by_i \ll \lambda z$, Eq. (3.11) reduces to

$$U(x_i, y_i) = ab \exp[i\alpha\Delta x_0 + i\beta\Delta y_0] \exp\left[i\left(\frac{2\pi x_i}{\lambda z}\right)\Delta x_0 + \frac{2\pi y_i}{\lambda z}\Delta y_0\right] \tag{3.12}$$

If, then, the computed complex amplitude of the object wave at a point in the hologram plane can be written as

$$o(n\Delta x_0, m\Delta y_0) = |o(n\Delta x_0, m\Delta y_0)| \exp[i\varphi(n\Delta x_0, m\Delta y_0)] \tag{3.13}$$

Its modulus and phase at this point can be encoded, by making the area of the opening located in this cell equal to the modulus so that

$$ab = | o(n\Delta x_0, m\Delta y_0) | \tag{3.14}$$

and displacing the center of the opening from the center of the cell by an amount δx_{mm} given by the relation

$$\delta x_{mm} = (\Delta x_0 / 2\pi)\ \varphi(n\Delta x_0, m\Delta y_0) \tag{3.15}$$

The total diffracted amplitude in the far field, which is obtained by summing the complex amplitudes due to all the $N \times N$ openings, considering the dimensions of the cells and the angle of illumination are chosen ($\alpha\Delta x_0 = 2\pi$, $\alpha\Delta y_0 = 2\pi\delta x_{mm} \ll \lambda z$), Eq. (3.12) can be written as

$$U(x_i, y_i) = \sum_{n=1}^{N}\sum_{n=1}^{N} | o(n\Delta x_0, m\Delta y_0) | \exp[i\Phi(n\Delta x_0, m\Delta y_0)]\exp[i\frac{2\pi}{\lambda z}(nx_i\Delta x_0 + my_i\Delta y_0)] \tag{3.16}$$

which is the discrete Fourier transform of the computed complex amplitude in the hologram plane or, in other words, the described reconstructed image.

Binary detour-phase holograms have several attractive features. It is possible to use a simple pen-and-ink plotter to prepare the binary master, and the problems of linearity do not arise in the photographic reduction process. Their chief disadvantage is that they are very wasteful of plotter resolution, since the number of addressable plotter points in each cell must be large to minimize the noise due to quantization of the modulus and the phase of the Fourier coefficients. When the number of phase-quantization levels is large, this noise is effectively spread over the whole image field, independent of the form of the signal. However, when the number of phase-quantization levels is small, the noise terms become shifted and self-convolved version of the signal, which are much more annoying.

(2) Computer-generated interferograms

Problems can arise with detour-phase holograms when encoding wavefronts with large phase variations, since a pair of apertures near the crossover may overlap when the phase of the wavefront moves through a multiple of $2n$ radians. This difficulty has been avoided in an alternative approach to the production of binary holograms based on the fact that an image hologram of a wavefront that has no amplitude variations is essentially similar to an interferogram, so that the exact locations of the transparent elements in the binary hologram can be determined by solving a grating equation.

Different methods can then be used to incorporate information on the amplitude variation in the object wavefront into the binary fringe pattern. In one method, the two-dimensional nature of the Fourier transform hologram is used to record the phase information along x direction, and the fringe heights in the y direction are adjusted to correspond to the amplitude. In another, the phase and the amplitude are recorded through the position and the width of the fringes along the direction of the carrier frequency, and in the third, the phase and amplitude of the object wave are encoded by the superimposition of two phase-only holograms.

(3) Optical application of CGHs

Now the CGHs have been widely used in daily life. One of the main applications of CGHs is interferometric tests of aspheric optical surfaces. Normally, such tests would require either an aspheric reference surface or an additional optical element, commonly referred to as a null lens, which converts the wavefront produced by the element under test into a spherical or plane wavefront.

An optical system using a Twyman-Green interferometer in conjunction with a CGH to test an aspheric mirror is shown in Figure 3-6. The hologram is a binary representation of the interferogram that would be obtained if the wavefront from an ideal aspheric surface were to interfere with a titled plane wavefront, and is placed in the plane in which the mirror under test is imaged. The superimposition of the actual interference fringes and the CGH produces a moire pattern which maps the deviation of the actual wavefront from the ideal computed wavefront.

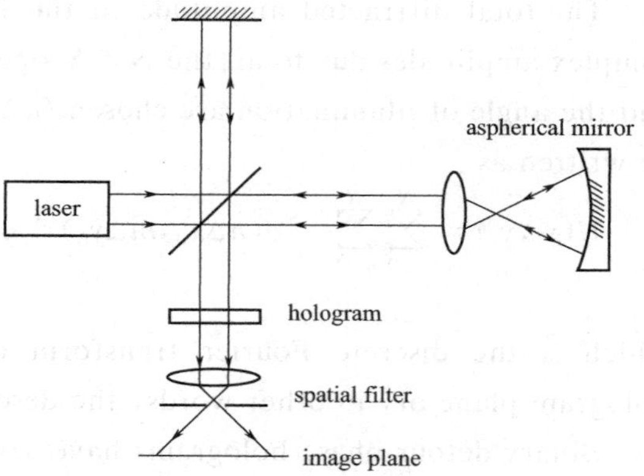

Figure 3-6 Twyman-Green interferometer modified to use a CGH to test an aspheric mirror

The contrast of the moire pattern is improved by spatial filtering. This is done by reimaging the hologram through a small aperture placed in the focal plane of the reimaging lens. The position of this aperture is chosen so that it passes only the transmitted wavefront from the mirror under test and the diffracted wavefront produced by illuminating the hologram with the plane reference wavefront. These two wavefronts can be isolated if, in producing the computer-generated hologram, the slope of the aspheric wavefront along the same direction. Typical fringe patterns obtained with an aspheric surface, with and without a computer-generated hologram, are shown in Figure 3-7. Interference patterns obtained with an aspheric wavefront having a maximum slope of 35 waves per radius and a maximum departure of 19 waves from a reference sphere.

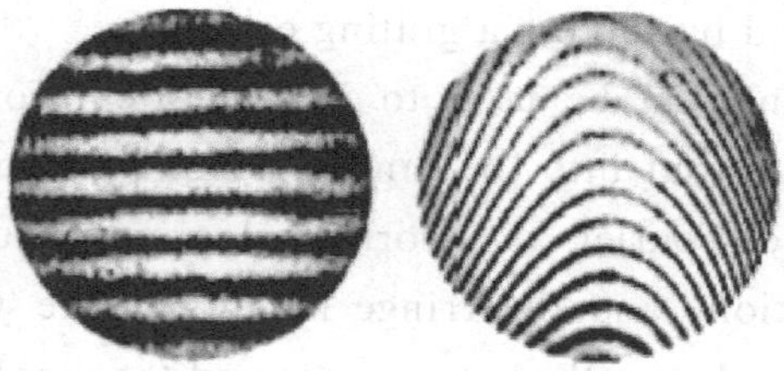

(a) with a computer-generated hologram (b) without a computer-generated hologram

Figure 3-7 Comparison of interference patterns

Computer-generated holograms have been widely used to test aspheric surfaces. With the development of improved plotting routines and the application of techniques such as electron-beam recording and using layers of photoresist coated on optically worked substrates for the production of computer-generated holograms of very high quality, recent work has focused on refinements of the technique to obtain the highest precision. A preferred interferometer configuration is one in which both test and reference beams pass through the CGH so that aberrations of the substrate have no significant effect on the interferogram. It is also essential that the beams incident on the CGH should be collimated in order to reduce the effects of misalignment. Other factors to be considered are the use of test surface images on CGH and the design of CGH, using ray-tracing software, to compensate for off-axis aberrations introduced by the imaging system.

3.2 Wave-Optics Analysis of Optical Systems

3.2.1 Lens as a Phase Transformation

A lens is composed of an optically dense material, usually glass with a refractive index of approximately 1.5, in which the propagation velocity of an optical disturbance is less than the velocity in air. A lens is said to be a thin lens if a ray entering at coordinates (x, y) on one face exits at approximately the same coordinates on the opposite face, i.e., if there is negligible translation of a ray within the lens. Thus a thin lens simply delays an incident wavefront by an amount proportional to the thickness of the lens at each point.

Referring to Figure 3-8, let the maximum thickness of the lens (on its axis) be Δ_0, and let the thickness at coordinates (x, y) be $\Delta(x, y)$. Then the total phase delay suffered by the wave at coordinates (x, y) in passing through the lens may be written.

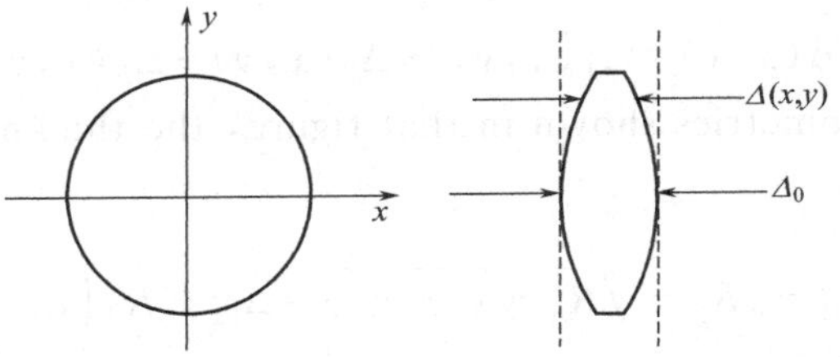

Figure 3-8 The thickness function

$$\Phi(x, y) = kn\Delta(x, y) + k[\Delta_0 - \Delta(x, y)] \tag{3.17}$$

where n is the refractive index of the lens material, $kn\Delta(x, y)$ is the phase delay introduced by the lens, and $k[\Delta_0 - \Delta(x, y)]$ is the phase delay introduced by the remaining region of free space between the two planes. Equivalently the lens may be represented by a multiplicative phase transformation of the form

$$t_l(x, y) = \exp(ik\Delta_0)\exp[ik(n-1)\Delta(x, y)] \tag{3.18}$$

The complex field $U_1'(x, y)$ across a plane immediately behind the lens is then related to the complex field $U_1(x, y)$ incident on a plane immediately in front of the lens by

$$U_1'(x, y) = t_1(x, y) U_1(x, y) \tag{3.19}$$

The problem remains to find the mathematical form of the thickness function $\Delta(x, y)$ in order that the effects of the lens may be understood.

(1) The thickness function

In order to specify the forms of the phase transformations introduced by a variety of different types of lenses, we first adopt a sign convention, i. e., as rays travel from left to right, each convex surface encountered is taken to have a positive radius of curvature, while each concave surface is taken to have a negative radius of curvature. Thus, in Figure 3-9 (b) the radius of curvature of the left-hand surface of the lens is a positive number R_1, while the radius of curvature of the right-hand surface is a negative number R_2.

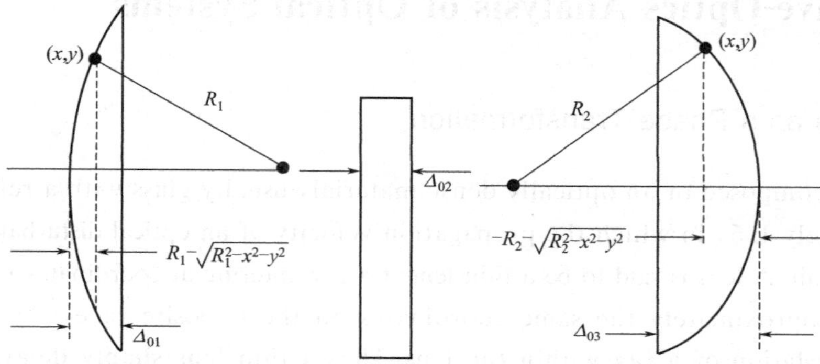

(a) geometry for Δ_1 (b) geometry for Δ_2 (c) geometry for Δ_3

Figure 3-9 Calculation of the thickness function

To find the thickness $\Delta(x, y)$, we split the lens into three parts, as shown in Figure 3-9, and write the total thickness function as the sum of three individual thickness functions.

$$\Delta(x, y) = \Delta_1(x, y) + \Delta_2(x, y) + \Delta_3(x, y) \tag{3.20}$$

Referring to the geometries shown in that figure, the thickness function $\Delta_1(x, y)$ is given by

$$\Delta_1(x, y) = \Delta_{01} - (R_1 - \sqrt{R_1^2 - x^2 - y^2}) = \Delta_{01} - R_1\left(1 - \sqrt{1 - \frac{x^2 + y^2}{R_1^2}}\right) \tag{3.21}$$

The second component of the thickness function comes from a region of glass of constant thickness Δ_{02}. The third component is given by

$$\Delta_3(x, y) = \Delta_{03} - (-R_2 - \sqrt{R_2^2 - x^2 - y^2}) = \Delta_{03} + R_2\left(1 - \sqrt{1 - \frac{x^2 + y^2}{R_2^2}}\right) \tag{3.22}$$

where we have factored the positive number—R_2 out of the square root. Combining the three expressions for thickness, the total thickness is seen to be

$$\Delta(x, y) = \Delta_0 - R_1\left(1 - \sqrt{1 - \frac{x^2 + y^2}{R_1^2}}\right) + R_2\left(1 - \sqrt{1 - \frac{x^2 + y^2}{R_2^2}}\right) \tag{3.23}$$

where $\Delta_0 = \Delta_{01} + \Delta_{02} + \Delta_{03}$.

(2) The paraxial approximation

The expression for the thickness function can be substantially simplified if attention is restricted to portions of the wavefront that lie near the lens axis, or equivalently, if only paraxial rays are considered. Thus we consider only values of x and y sufficiently small to allow the following approximations to be accurate

$$\sqrt{1-\frac{x^2+y^2}{R_1^2}} \approx 1-\frac{x^2+y^2}{2R_1^2}, \quad \sqrt{1-\frac{x^2+y^2}{R_2^2}} \approx 1-\frac{x^2+y^2}{2R_2^2} \tag{3.24}$$

The resulting phase transformation will, of course, represent the lens accurately over only a limited area, but this limitation is no more restrictive than the usual paraxial approximation of geometrical optics. Note that Eq. (3.24) amounts to approximations of the spherical surfaces of the lens by parabolic surfaces. With the help of these approximations, the thickness function becomes

$$\Delta(x, y) = \Delta_0 - \frac{x^2+y^2}{2}\left(\frac{1}{R_1}-\frac{1}{R_2}\right) \tag{3.25}$$

(3) The phase transformation and its physical meaning

Substitution of Eq. (3.25) into Eq. (3.18) yields the following approximation to the lens transformation

$$t_1(x, y) = \exp(ikn\Delta_0)\exp\left[-ik(n-1)\frac{x^2+y^2}{2}\left(\frac{1}{R_1}-\frac{1}{R_2}\right)\right] \tag{3.26}$$

The physical properties of the lens (that is, n, R_1 and R_2) can be combined in a single number f called the focal length, which is defined by

$$\frac{1}{f} = (n-1)\left(\frac{1}{R_1}-\frac{1}{R_2}\right) \tag{3.27}$$

Neglecting the constant phase factor, which we shall drop hereafter, the phase transformation may now be rewritten

$$t_1(x, y) = \exp\left[-i\frac{k}{2f}(x^2+y^2)\right] \tag{3.28}$$

This equation will serve as our basic representation of the effects of a thin lens on an incident disturbance, where we neglect the finite extent of the lens.

The physical meaning of the lens transformation can best be understood by considering the effect of the lens on a normally incident, unit-amplitude plane wave. The field distribution U_1 in front of the lens is unity, Eq. (3.19) and Eq. (3.28) yield the following expression for U_1' behind the lens

$$U_1'(x, y) = \exp\left[-i\frac{k}{2f}(x^2+y^2)\right] \tag{3.29}$$

3.2.2 Frequency Analysis of Optical Imaging Systems

Considering the long and rich history of optics, the tools of frequency analysis and linear system theory have played important roles for only a relatively short period of time. Nevertheless, in this short time these tools have been so widely and successfully used that they now occupy a fundamental place in the theory of imaging systems.

(1) Generalized treatment of imaging systems

Suppose that an imaging system of interest is composed, not of a single thin lens, but perhaps of several lenses, some positive, some negative, with various distances between them. The lenses need not be thin in the sense defined earlier. We shall assume, however, that the system ultimately produces a real image in space; this is not a serious restriction, for if the system produces a virtual image, to view that image it must be converted to a real image, perhaps by the lens of the eye.

To specify the properties of the lens system, we adopt the point of view that all imaging elements may be lumped into a single "black box", and that the significant properties of the system can be completely described by specifying only the terminal properties of the aggregate. Referring to Figure 3-10, the "terminals" of this black box consist of the planes containing the entrance and exit pupils. It is assumed that the passage of light between the entrance pupil and the exit pupil is adequately described by geometrical optics.

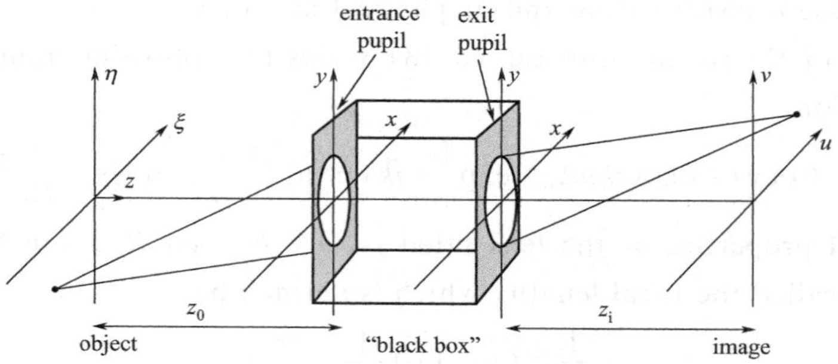

Figure 3-10 Generalized model of an imaging system

The entrance and exit pupils are in fact images of the same limiting aperture within the system. As a consequence, there are several different ways to visualize the origin of the spatial limitation of the wavefront that ultimately give rise to diffraction. It can be viewed as being caused by the physical limiting aperture internal to the system (which is the true physical source of the limitation). Equivalently it can be viewed as arising from the entrance pupil or from the exit pupil of the system.

We shall use the symbol z_0 to represent the distance of the plane of the entrance pupil from the object plane, and the symbol z_i to represent the distance of the plane of the exit pupil from the image plane. The distance z is then the distance that will appear in the diffraction equations that represent the effect of diffraction by the exit pupil on the point-spread function of the optical system. We shall refer either to the exit pupil or simply to the "pupil" of the system when discussing these effects.

An imaging system is said to be diffraction-limited if a diverging spherical wave, emanating from a point-source object, is converted by the system into a new wave, again perfectly spherical, that converges towards an ideal point in the image plane, where the location of that ideal image point is related to the location of the original object point through a simple scaling factor (the magnification), a factor that must be the same for all points in the image field of interest if the system is to be ideal. Thus the terminal property of a diffraction-limited imaging system is that a diverging spherical wave incident on the entrance pupil is converted by the system into a converging spherical wave at the exit pupil. For any real imaging system, this property will be satisfied, at best, over only finite regions of the object and image planes. If the object of interest is confined to the region for which this property holds, then the system may be regarded as being diffraction-limited.

Since geometrical optics adequately describes the passage of light between the entrance and exit pupils of a system, diffraction effects play a role only during passage of light from the object to the entrance pupil, or alternatively and equivalently, from the exit pupil to the image. It is, in fact, possible to associate all diffraction limitations with either of these two pupils. The two points of view that regard image resolution as being limited by: ① the finite entrance pupil seen from the object space or ② the finite exit pupil seen from the image space is entirely equivalent, due to the fact that these two pupils are images of each other.

(2) Frequency response for diffraction-limited coherent imaging

A coherent imaging system is linear in complex amplitude. This implies, of course, that such a system provides a highly nonlinear intensity mapping. If frequency analysis is to be applied in its usual form, it must be applied to the linear amplitude mapping. We would anticipate the transfer function concepts applied directly to this system, provided it is done on an amplitude basis. To do so, define the following frequency spectra of the input and output, respectively

$$G_g(f_x, f_y) = \iint_{-\infty}^{+\infty} U_g(u, v)\exp[-i2\pi(f_x u + f_y v)]\mathrm{d}u\mathrm{d}v \tag{3.30}$$

$$G_i(f_x, f_y) = \iint_{-\infty}^{+\infty} U_i(u, v)\exp[-i2\pi(f_x u + f_y v)]\mathrm{d}u\mathrm{d}v \tag{3.31}$$

In addition, define the amplitude transfer function H as the Fourier transform of the space invariant amplitude impulse response

$$H(f_x, f_y) = \iint h(u, v)\exp[-i2\pi(f_x u + f_y v)]\mathrm{d}u\mathrm{d}v \tag{3.32}$$

Hence
$$G_i(f_x, f_y) = H(f_x, f_y)\, G_g(f_x, f_y) \tag{3.33}$$

Thus the effects of the diffraction-limited imaging system have been expressed, at least formally, in the frequency domain. It now remains to relate H more directly to the physical characteristics of the imaging system itself.

Finally we give some intuitive explanation as to why the scaled pupil function plays the role of the amplitude transfer function. Remember that in order to completely remove the quadratic phase factor across the object, the object should be illuminated with a spherical wave, in this case converging towards the point where the entrance pupil is pierced by the optical axis. The converging spherical illumination causes the Fourier components of the object amplitude transmittance to appear in the entrance pupil, as well as in the exit pupil, since the latter is the image of the former. Thus the pupil sharply limits the range of

Fourier components passed by the system.

(3) Frequency response for diffraction-limited incoherent imaging

In the coherent case, the relation between the pupil and the amplitude transfer function has been seen to be a very direct and simple one. When the object illumination is incoherent, the transfer function of the imaging system will be seen to be determined by the pupil again, but in a less direct and somewhat more interesting way. The theory of imaging with incoherent light has, therefore, a certain extra richness not present in the coherent case. We turn now to considering this theory; again attention will be centered on diffraction-limited systems, although the discussion that immediately follows applies to all incoherent systems, regardless of their aberrations.

Imaging systems that use incoherent illumination have been seen to obey the intensity convolution integral

$$I_i(u,v) = K\iint |h(u-\tilde{\xi}, v-\tilde{\eta})|^2 I_g(\xi,\eta)\mathrm{d}\tilde{\xi}\mathrm{d}\tilde{\eta} \tag{3.34}$$

Such systems should therefore be frequency-analyzed as linear mappings of intensity distributions. To this end, let the normalized frequency spectra of I, and I_i be defined by

$$G_g(f_x,f_y) = \frac{\iint I_g(u,v)\exp[-i2\pi(f_x u + f_y v)\mathrm{d}u\mathrm{d}v]}{\iint I_g(u,v)\mathrm{d}u\mathrm{d}v} \tag{3.35}$$

$$G_i(f_x,f_y) = \frac{\iint I_i(u,v)\exp[-i2\pi(f_x u + f_y v)\mathrm{d}u\mathrm{d}v]}{\iint I_i(u,v)\mathrm{d}u\mathrm{d}v} \tag{3.36}$$

The normalization of the spectra by their "zero-frequency" values is partly for mathematical convenience, and partly for a more fundamental reason. It can be shown that any real and nonnegative function, such as I or I_i, has a Fourier transform which achieves its maximum value at the origin. We choose that maximum value as a normalization constant in defining G_g and G_i. Since intensities are nonnegative quantities, they always have a spectrum that is nonzero at the origin. The visual quality of an image depends strongly on the "contrast" of the image, or the relative strengths of the information-bearing portions of the image and the ever-present background. Hence the spectra are normalized by that background.

In a similar fashion, the normalized transfer function of the system can be defined by

$$H(f_x,f_y) = \frac{\iint |h(u,v)|^2\exp[-i2\pi(f_x u + f_y v)\mathrm{d}u\mathrm{d}v]}{\iint |h(u,v)|^2\mathrm{d}u\mathrm{d}v} \tag{3.37}$$

Application of the convolution theorem to Eq. (3.34) then yields the frequency-domain relation

$$G_i(f_x,f_y) = H(f_x,f_y)\,G_g(f_x,f_y) \tag{3.38}$$

(4) Comparison of coherent and incoherent imaging

As seen in previous sections, the optical transfer function (OTF) of a diffraction-limited system extends to a frequency that is twice the cutoff frequency of the amplitude transfer function. It is tempting, therefore, to conclude that incoherent illumination will invariably yield "better" resolution than coherent illumination, given that the same imaging system is used in both cases. As we shall now see, this conclusion is in general not a valid one; a comparison of the two types of illumination is far more complex than such a superficial examination would suggest.

A major flaw in the above argument lies in the direct comparison of the cutoff frequencies in the two cases. Actually, the two are not directly comparable, since the cutoff of the amplitude transfer function determines the maximum frequency component of the image amplitude while the cutoff of the optical transfer function determines the maximum frequency component of image intensity. Surely any direct comparison of the two systems must be in terms of the same observable quantity, image intensity.

In the absence of a meaningful quality criterion, we can only examine certain limited aspects of the two types of images, realizing that the comparisons so made will probably bear little direct relation to overall image quality. Nonetheless, such comparisons are highly instructive, for they point out certain fundamental differences between the two types of illumination.

3.3 Optical Processing

Optical processing and related areas are important branches of modern optics and cannot easily be done justice in a few pages. Here we offer an introduction and concentrate as usual on a physical understanding of what can sometimes require enormously complicated mathematical treatments.

We begin by considering the standard optical processor of Figure 3-11. The object is illuminated by a coherent plane wave. Two identical lenses are placed in the locations shown in the figure. A simple, paraxial ray trace will show that the lenses project an inverted image into the focal plane of the second lens. Further, the principal planes lying between the two lenses are equidistant from object and image. The magnification of the processor is therefore 1.

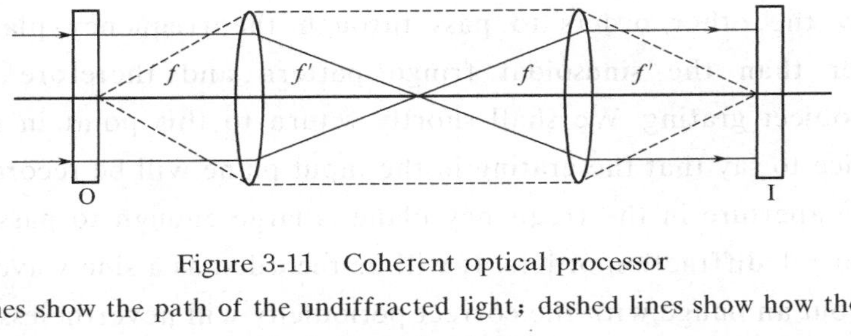

Figure 3-11 Coherent optical processor

Solid lines show the path of the undiffracted light; dashed lines show how the image is projected by the lenses

Thus, the processor projects a real, inverted image at unit magnification.

3.3.1 Abbe Theory

We now approach the processor from the point view of wave optics. The treatment that follows is identical to parts of the Abbe theory of microscopy, and we use the term, optical processing, to include many of the techniques of microscopy.

To simplify the explanation, the object is assumed to be a grating whose period is d, located in the input plane. The focal plane of the first lens is known as the frequency plane for reasons that will become clear later. We can find the distribution of intensity in the frequency plane by applying the grating equation

$$m\lambda = d\sin\theta \qquad (m=0,\ \pm 1,\ \pm 2,\cdots) \tag{3.39}$$

In paraxial approximation, $\sin\theta \doteq \theta$, and principal maxima are located in the frequency plane at positions

$$\xi = m\lambda f'/d \tag{3.40}$$

where $\xi = 0$ corresponds to the intersection of the optical axis with frequency plane.

Consider, for the moment, any two adjacent diffraction orders, such as $m = 0, +1$, assume that all other orders have somehow been suppressed. The second lens is then presented with two point sources a distance ξ apart. The intensity distribution in the output plane is, apart from a multiplicative constant

$$I' = \cos^2[(\pi/\lambda)\xi\sin\theta'] \tag{3.41}$$

where the prime refers to the output (image) plane. In paraxial approximation

$$\theta' = x'/f' \tag{3.42}$$

so that

$$I'(x') = \cos^2\left(\frac{\pi}{\lambda}\frac{\xi x'}{f'}\right) \tag{3.43}$$

The spacing d' between adjacent maxima of the fringes is found from the relation

$$\frac{\pi}{\lambda}\frac{\xi d'}{f'} = \pi \tag{3.44}$$

The value of ξ is $\lambda f'/d$ (since $m = 1$). Thus,

$$d' = d \tag{3.45}$$

The grating in the input plane appears in the output plane as a sinusoidal interference pattern with precisely the same periodicity as the original grating.

If we allow the other orders to pass through the frequency plane, the grating becomes sharper than the sinusoidal fringe pattern and therefore more faithfully reproduces the object grating. We shall shortly return to this point in more detail. For now, let it suffice to say that the grating in the input plane will be recorded in the image plane only if the aperture in the frequency plane is large enough to pass both the 0 and either the +1 or −1 diffraction orders. It will be recorded as a sine wave, rather than as a square wave, but an image with the correct periodicity will nevertheless be detectable in the output plane.

Until now, we have tacitly assumed that the object is an amplitude transmission grating, consisting of alternately clear and opaque strips. We may take a glimpse of the power of optical processing if we allow the object to be a phase grating—that is, one that is transparent, but whose optical thickness varies periodically with the dimension x of the input plane. Ordinarily, the object is nearly invisible, because it is wholly transparent. Nevertheless, it exhibits diffracted orders, and one way (though not the best way) to render it visible would be to locate a special screen or spatial filter in the frequency plane and permit only the orders +1 and 0 to pass through holes in the screen. Then, as before, the image would be easily visible as a sinusoidal fringe pattern with the proper spacing between maxima. Spatial filtering is thus able to transform objects that are substantially invisible into visible images. This is the principle of phase microscopy, which is treated in more detail subsequently.

3.3.2 Fourier-Transform Optics

The amplitude distribution in the frequency plane is then proportional to the Fourier transform $G(fx)$. If the object is complicated and is characterized by an amplitude transmittance function $g(x)$, we must generalize our earlier Fraunhofer diffraction integral to include a factor $g(x)$ in the integrand

$$E(\theta) = \frac{iA}{\lambda}\int_{-b/2}^{b/2} g(x)\mathrm{e}^{-ik\sin\theta x}\mathrm{d}x \tag{3.46}$$

or, in terms of the dimension ξ in the frequency plane

$$E(\xi) = \int_{-b/2}^{b/2} g(x)\mathrm{e}^{-i\frac{2\pi\xi x}{\lambda f'}}\mathrm{d}x \tag{3.47}$$

apart from the multiplicative constants. If we define a new variable f_x

$$f_x = \frac{\xi}{\lambda f'} \tag{3.48}$$

We have

$$E(f_x) = \int_{-b/2}^{b/2} g(x)\mathrm{e}^{-i2\pi f_x x}\mathrm{d}x \tag{3.49}$$

where $g(x)$ is a mathematical function that describes the object in the input plane and $E(f_x)$ is a representation of the electric-field amplitude in the frequency plane.

$E(f_x)$ greatly resembles the Fourier transform of $g(x)$. To make the resemblance perfect, we have only to define $g(x)$ to be zero outside the range $-b/2<x<b/2$ and extend the range of integration from $-\infty$ to $+\infty$. The amplitude distribution in the frequency plane is then proportional to the Fourier transform $G(f_x)$ of $g(x)$

$$G(f_x) = \int_{-\infty}^{+\infty} g(x)\mathrm{e}^{-i2\pi f_x x}\mathrm{d}x \tag{3.50}$$

We know from geometric optics that the processor casts an inverted image into the output plane. Thus, we define the positive x'' axis to have opposite direction from the positive x' axis. When the x' axis is defined in this way, the amplitude distribution $g'(x')$ in the output plane will be identical to that in the input plane.

The theory of the Fourier transform shows that the inverse transform

$$g'(x') = \int_{-\infty}^{+\infty} G(f_x)\mathrm{e}^{-i2\pi x' f_x}\mathrm{d}f_x \tag{3.51}$$

is also equal to $g(x)$. We thus conclude that the second lens performs the inverse transform, provided only that the x axis be defined, as above, to take into account the fact that the system projects an inverted image. Needless to say, the fact can also be derived by rigorous mathematics.

3. 3. 3 Spatial Filtering

This term is usually used to describe manipulation of an image with masks in the frequency plane. We have already encountered some examples in connection with the Abbe theory and the Fourier series.

The simplest kind of spatial filter is a pinhole located in the focal plane of a lens. It acts as a low-pass filter and is commonly used to improve the appearance of gas-laser beams.

A gas-laser beam is typically highly coherent. The presence of small imperfection in a microscope objective, for example, results in a certain amount of scattered light. In an incoherent optical system this is of minor importance. Unfortunately, the scattered light in a coherent system interferes with the unscattered light to produce unsightly ring patterns that greatly resemble Fresnel zone plates.

Fortunately, the rings have relatively high spatial frequencies, so that these frequencies can be blocked by focusing the beam through a hole that transmits nearly the entire beam. The hole should be a few times the diameter of the Airy disk, so that very little other than the scattered light is lost.

Another important spatial filter is the high-pass filter. This consists of a small opaque spot in the center of the frequency plane. The spot blocks the low-frequency components of the object's spatial-frequency spectrum and allows the high-frequency components to pass. High-pass filtering can be used to sharpen photographs or to aid in examining fine detail.

Spatial filtering may also be used to remove unwanted detail from a photograph or to identify a character or a defect in a photograph. For example, suppose for some reasons that we had taken a photograph of a video display. The picture is composed of approximately 500 discrete, horizontal lines. The spacing of the lines determines the highest spatial frequency in the photograph. If we place the photograph in the input plane of the processor, we will see very strong diffraction orders in the frequency plane. These orders correspond to the harmonics of the grating formed by the horizontal lines.

To eliminate the lines in the output plane, we carefully insert two knife edges in the frequency plane. We locate the knife edges so that they cut off the +1 and −1 diffraction orders of the grating but pass all lower spatial frequencies. The result is that the picture is passed virtually unchanged, but the lines are eliminated completely. The picture has not been blurred, and the finest details are still visible in the output plane. Only the lines are absent.

New Words and Expressions

a glimpse of　瞥见
adjacent　邻近的
aggregate　集合
aspheric　非球面的
attenuate　衰减
convention　法则，习俗
degrade　使降级，使变差
encounter　遭遇
etalon　校准器
equivalently　等价地，相当于
factor　分解
match　匹配
microscope objective　显微物镜
proportional　成比例的
provided only　只要
quantization　量化
refinement　精加工
sampling theorem　抽样定理
scattered light　散射光
transmittance　透过率
transparency　透明度，透光度
tungsten lamp　钨丝灯
wavefront reconstruction　波前重构

Translation

1. In Gabor's original holography, light was filtered and passed through a pinhole to bring about the necessary coherence.

翻译：在盖博最初的全息技术中，光被过滤并通过一个针孔来产生干涉。

2. In addition, if the thickness of the recording medium is large compared to the fringe spacing, volume effects are important.

翻译：另外，如果记录介质的厚度比条纹间距大，体积效应就很重要。

3. The resulting recording can be considered as made up of three incoherently superposed holograms.

翻译：由此产生的记录可以认为是由三个不相干叠加的全息图组成的。

4. After those three light expose the recording holograms for three times respectively, the rainbow hologram can be formed in the dry plate.

翻译：这三种光分别曝光并记录全息图三次后，在干板上形成彩虹全息图。

5. Several methods have been tried to eliminate these cross-talk images, including spatial-frequency multiplexing, spatial multiplexing or coded reference waves, and division of the aperture field.

翻译：有多种方法应用于消除这些串扰图像，包括空间频率复用、空间复用或编码参考波以及孔径场的划分。

6. The first method to eliminate cross-talk that did not involve such penalties was based on the use of volume holograms.

翻译：消除串扰的第一种方法是基于体全息图的使用。

7. This principle was first applied to produce a two-color hologram of a transparency, and subsequently extended to three-color imaging of diffusely reflecting objects.

翻译：这一原理首先被用于产生一个透明的双色全息图，随后扩展到扩散反射物体的三色成像。

8. When this hologram was illuminated once again with a similar multicolor beam at the appropriate angle, a multicolor reconstructed image with negligible cross-talk was obtained.

翻译： 当用类似的多色光束以适当的角度再次照射该全息图时，获得了可忽略串扰的多色再现像。

9. This made it possible to produce holograms that reconstruct very bright multicolor images when illuminated with a white-light source.

翻译： 这使得当用白色光源照明时，可以产生再现非常明亮的多色图像的全息图。

10. The object, turned sideways, was placed on one side of the axis of the mirror so that its image was formed on the other side, at the same distance from the mirror.

翻译： 将物体侧向转动，放在镜子轴的一边，这样它的影像就形成在另一边，与镜子的距离相同。

11. One of the most popular approaches to generate CGH involves two principal steps.

翻译： 生成计算全息最流行的方法之一包括两个主要步骤。

12. Such a hologram has the advantage that it forms a single, on-axis image.

翻译： 这种全息图的优点是它会形成一个单一的轴上图像。

Unit 4
Electromagnetic Fields and Electromagnetic Waves

4.1 The Concept of Electromagnetic Fields and Waves

EM field can be viewed as the combination of an electric field and a magnetic field. As we know, a static charge gives rise to an effect that appears as a force acting on a charge body in the surrounding, within which an electric field is said to exist. Moving charges or electric currents lead to another kind of field, which results in forces acting on magnets and conductors carrying currents, referred to as a magnetic field.

The electric field and the magnetic field are both the vector fields. If the magnitude and the location of the electric charge do not change with time, the electric field produced by the charge will also be constant over time. This kind of electric field is known as an electrostatic field. When the magnitude and the velocity of an electric charge in motion are kept constant so that the resultant electric current is steady, the magnetic field produced will be time independent and known as a steady magnetic. If the charge and the current vary with time, the electric field and the magnetic field they produced will be a function of time. It was found that time-varying electric field and magnetic field must co-exist and have definite relation to each other, leading to a time-varying EM field. The EM field extends indefinitely throughout space and describes the EM interaction.

EM wave (radiation) is a self-propagating wave in space with electric and magnetic components. It consists of discrete packets of energy, which we call photons. A photon consists of an oscillating electric field component, E, and an oscillating magnetic field component, M. A moving charge gives rise to a magnetic field, and if the motion is changing (accelerated), then the magnetic field varies and in turn produces an electric field. These interacting electric and magnetic fields are at right angles to one another and also to the direction of propagation of the energy. That is, the electric and magnetic fields are orthogonal (perpendicular) to each other, and they are orthogonal to the direction of propagation of the photon. Thus, an EM wave is a transverse wave. If the direction of the electric field is constant, the wave is said to be polarized. EM radiation does not require a material medium and can travel through a vacuum. The electric and magnetic fields of a photon flip direction as the photon travels. We call the number of flips, or oscillations, that occur in one second the frequency, v.

From a classical perspective, the EM field can be regarded as a smooth, continuous field, propagated in a wavelike manner; whereas, from a quantum mechanical perspective, the field is seen as quantized, being composed of individual particles. Electric charges and currents are sources for EM fields. It should be pointed out that they are the only sources for producing EM fields. Up to now, no magnetic charge or magnetic current of significance has been found exist in nature. The way in which charges and currents interact with the EM field is described by Maxwell's equations.

4.2 The History of Electromagnetic Waves

Electric and magnetic phenomena have been known for millenia. The earliest examples were the forces produced by static electricity and by ferromagnetism. Of course there are other phenomena that we recognize today to be of EM origin, which have been observed since the beginning of time. For example, lighting is an electric discharge. Also, light consists of EM waves, in the quantum theory, photons. But before the scientific revolution, it was not recognized that these varied phenomena have a common origin.

The quantitative study of electricity and magnetism began with the scientific research of the French physicist Charles Augustin Coulomb. In 1787 Coulomb proposed a law of force for charges that, like Sir Isaac Newton's law of gravitation, varied inversely as the square of the distance. Using a sensitive torsion balance , he demonstrated its validity experimentally for forces of both repulsion and attraction. Like the law of gravitation, Coulomb's law was based on the notion of "action at a distance", wherein bodies can interact instantaneously and directly with one another without the intervention of any intermediary.

At the beginning of the nineteenth century, the electrochemical cell was invented by Alessandro Volta, professor of natural philosophy at the University of Pavia in Italy. The cell created an electromotive force , which made the production of continuous currents possible.

In 1820 at the University of Copenhagen, Hans Christian Oersted, professor of physics, made the momentous discovery that an electric current in a wire could deflect a magnetic needle. This experiment illustrated the connection between electricity and magnetism. News of this discovery was communicated to the French Academy of Sciences two months later. These results sparked experiments across the globe as scientists attempted to find an explanation. The laws of force between current bearing wires were at once investigated by Andre-Marie Ampere and by Jean-Baptiste Biot and Felix Savart. The favoured explanation, which was provided by Ampere, was that central forces were the cause. Within six years the theory of steady currents was complete.

Subsequently, in 1831, the British scientist Michael Faraday demonstrated the reciprocal effect, in which a moving magnet in the vicinity of a coil of wire produced an electric current. This phenomenon, together with Oersted's experiment with the magnetic needle, led Faraday to conceive the notion of a magnetic field. A field produced by a current in a wire interacted with a magnet. Also, according to his law of induction, a time varying magnetic field incident on a wire would induce a voltage, thereby creating a current. Electric forces could similarly be expressed in terms of an electric field created by the presence of a charge.

By 1850 Faraday had completed much of his work but he did not formulate his laws mathematically and the majority of scientists had failed to realize its significance. It was left to the Scottish physicist James Clerk Maxwell to establish the mathematical theory of electromagnetism based on the physical concepts of Faraday. In a series of papers published between 1856 and 1865 , Maxwell restated the laws of Coulomb, Ampere, and Faraday in terms of Faraday's electric and magnetic fields. Maxwell thus unified the theories of electricity and magnetism, in the same sense that two hundred years earlier Newton had unified terrestrial and celestial mechanics through his theory of universal gravitation.

As is typical of abstract mathematical reasoning, Maxwell saw in his equations a certain symmetry that suggested the need for an additional term, involving the time rate of change of the electric field. With this generalization, Maxwell's equations also became consistent with the principle of conservation of charge. Furthermore, Maxwell made the profound observation that his set of equations, thus modified, predicted the existence of EM waves. Therefore, disturbances in the EM field could propagate through space. Using the values of known experimental constants obtained solely from measurements of charges and currents, Maxwell deduced that the speed of propagation was equal to speed of light. This quantity had been measured astronomically by Olaf Romer in 1676 from the eclipses of Jupiter' s satellites and determined experimentally from terrestrial measurements by H. L. Fizeau in 1849. He then asserted that light itself was an EM wave, thereby unifying optics with electromagnetism as well.

Maxwell's theory was not accepted by scientists immediately, in part because it had been derived from a bewildering collection of mechanical analogies and difficult mathematical concepts. The form of Maxwell's equations as we are known today is due to the German physicist Heinrich Hertz. Hertz simplified them and eliminated unnecessary assumptions.

Hertz's interest in Maxwell's theory was occasioned by a prize offered by the Berlin Academy of Sciences in 1879 for research on the relation between polarization in insulators and EM induction. By means of his experiments, Hertz discovered how to generate high frequency electrical oscillations. He was surprised to find that these oscillations could be detected at large distances from the apparatus. Up to that time, it had been generally assumed that electrical forces decreased rapidly with distance according to the Newtonian law. He therefore sought to test Maxwell's prediction of the existence of EM waves.

In 1888 , Hertz set up standing EM waves using an oscillator and spark detector of his own design and made independent measurements of their wavelength and frequency. He found that their product was indeed the speed of light. He also verified that these waves behaved according to all the laws of reflection, refraction, and polarization that applied to visible light, thus demonstrating that they differed from light only in wavelength and frequency. "Certainly it is a fascinating idea," Hertz wrote, "that the processes in air that we have been investigating represent to us on a million-fold larger scale the same processes which go on in the neighborhood of a Fresnel mirror or between the glass plates used in exhibiting Newton's rings. "

On this foundation, around the 19th century, Popov in Russia and Marconi in Italy invented the technology to transmit information using EM waves, paving the way for the subsequent development of modern wireless communications, broadcasting, radar, remote control, microwave sensing, wireless networks and local area networks, satellite positioning optical communications and other information technologies. The wide applications of these new technologies further enhance the development of EM theory.

The availability of high performance and high speed computers and large memory capacity not only made the calculations encountered in obtaining the solutions to many problems in EMs possible, but also gave rise to new methods to compute EM fields and waves. This gave birth to computational EMs, which is an important branch of modern EMs.

4.3 The Basic Laws of Electromagnetic Theory

We know from experiments that charges, even though separated in vacuum, experience a mutual interaction. Recall the familiar electrostatics demonstration in which a pith ball somehow senses the presence of a charged rod without actually touching it. As a possible explanation we might speculate that each charge emits (and absorbs) a stream of undetected particles (virtual photons). The exchange of these particles among the charges may be regarded as the mode of interaction. Alternatively, we can take the classical approach and imagine instead that every charge is surrounded by something called an electric field. We then need only suppose that each charge interacts directly with the electric field in which it is immersed. Thus if a point charge q experiences a force F_E, the electric field E at the position of the charge is defined by $F_E = qE$. In addition, we observe that a moving charge may experience another force F_M, which is proportional to its velocity v. We are thus led to define yet another field, namely, the magnetic induction or just the magnetic field B, such that $F_M = qvB$. If force F_E and F_M occur concurrently, the charge is moving through a region pervaded by both electric and magnetic fields, whereupon $F = qE + qvB$.

As we'll see, electric fields are generated by both electric charges and time-varying magnetic fields. Similarly, magnetic fields are generated by electric currents and by time-varying electric fields. This interdependence of E and B is a key point in the description of light.

4.3.1 Faraday's Induction Law

"Convert magnetism into electricity" was the brief remark Michael Faraday jotted in his notebook in 1822, a challenge he set himself with an easy confidence that made it seem so attainable. After several years doing other research, Faraday returned to the problem of EM induction in 1831. His first apparatus made use of two coils mounted on a wooden spool, as shown in Figure 4-1 (a).

One, called the primary, was attached to a battery and a switch; the other, the secondary, was attached to a galvanometer. He found that the galvanometer deflected in one direction just for a moment whenever the switch was closed, returning to zero almost immediately, despite the constant current still in the primary. Whenever the switch was opened, interrupting the primary current, the galvanometer in the secondary circuit momentarily swung in the opposite direction and then promptly returned to zero.

Using a ferromagnetic core to concentrate the "magnetic force", Faraday wound two coils around opposing sections of a soft iron ring, as shown in Figure 4-1 (b). Now the effect was unmistakable—a changing magnetic field generated a current. Indeed, as he would continue to discover, change was the essential aspect of EM induction.

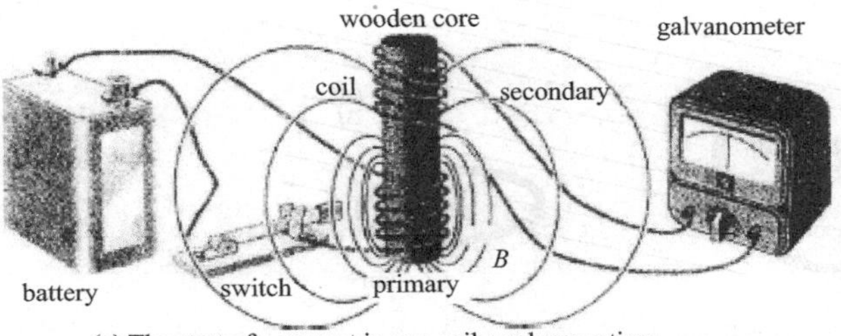

(a) The start of a current in one coil produces a time varying magnetic field that induces a current in the other coil

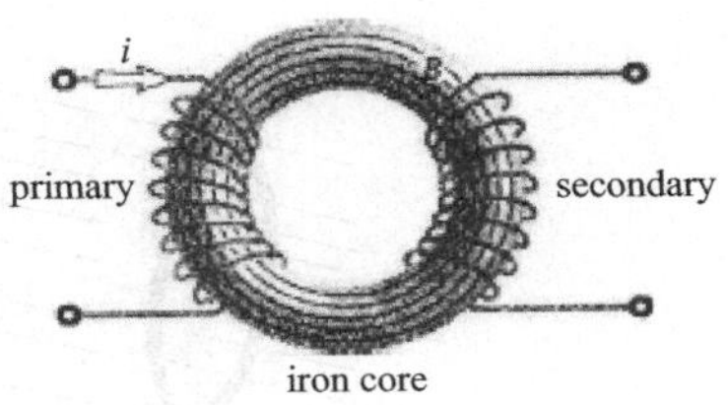

(b) An iron core couples the primary coil to the secondary

Figure 4-1 The schematic diagram of the Faraday's apparatus to produce EM induction

By thrusting a magnet into a coil, Faraday showed that there is a voltage—otherwise known as the induced electromotive force or emf—across the coil's terminals. (Electromotive force is a dreadful, outmoded term. It's not a force, but a voltage, so we'll avoid it and just use emf.) Furthermore, the amplitude of the emf depends on how rapidly the magnet is moved. The induced emf depends on the rate-of-change of B through the coil and not on B itself. A weak magnet moved rapidly can induce a greater emf than a strong magnet move slowly.

When the same changing B-field passes through two different wire loops, as in Figure 4-2, the induced emf is larger across terminals of the larger loop.

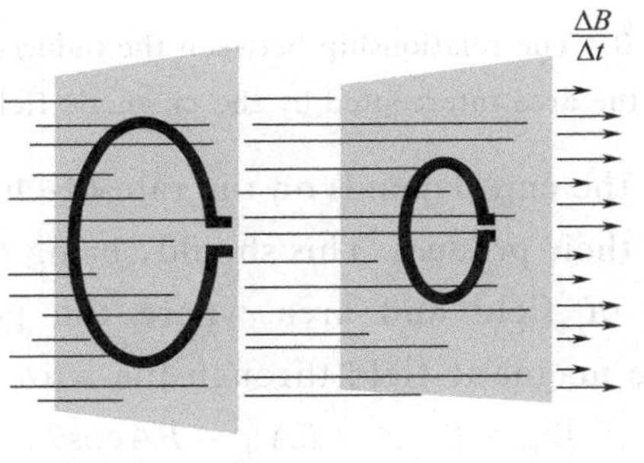

Figure 4-2 The larger time-varying magnetic flux passes through the larger coil and generates more electromotive force on its terminals

In other words, here where the field is changing, the induced emf is proportional to the area A of the loop penetrated perpendicularly by the field. If the loop is successively tilted over, as is shown in Figure 4-3, the area presented perpendicularly to the field ($A_{\perp}$) varies as $A\cos\theta$, and, when $\theta = 90^{\circ}$, the induced emf is zero because no amount of B-field then penetrates the loop: when $\Delta B/\Delta t \neq 0$, emf$\propto A_{\perp}$. The converse also holds: when the field is constant, the induced emf is proportional to the rate-of-change of the perpendicular area penetrated. If a coil is twisted or rotated or even squashed while in a constant B-field so that the perpendicular area initially penetrated is altered, there will be an induced emf$\propto A_{\perp}/\Delta t$ and it will be proportional to B. In summary, when $A_{\perp}$ = constant, emf $\propto A_{\perp} \Delta B/\Delta t$ and, when B = constant, emf $\propto B\Delta A_{\perp}/\Delta t$.

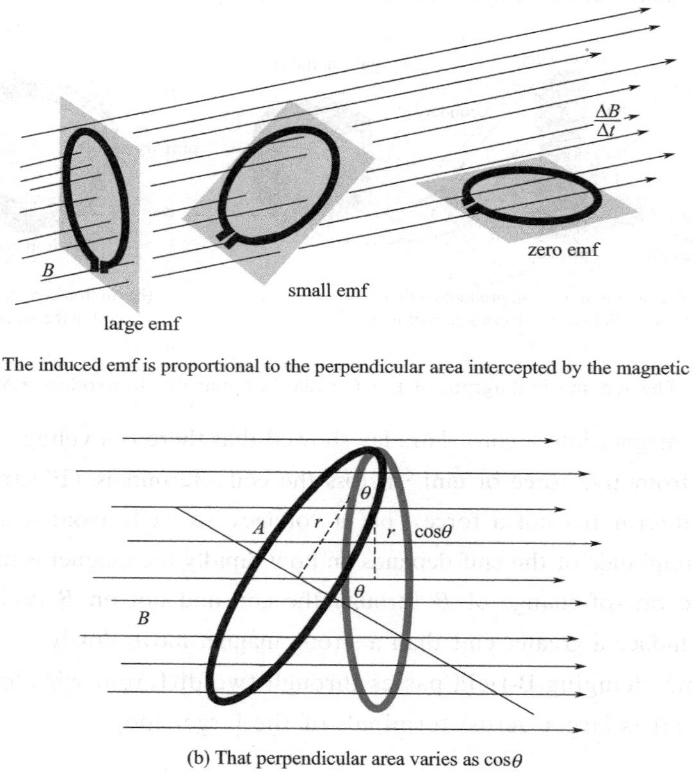

(a) The induced emf is proportional to the perpendicular area intercepted by the magnetic field

(b) That perpendicular area varies as cosθ

Figure 4-3 The relationship between the induced emf and the area intercepted by the magnetic field

All of this suggests that the emf depends on the rate-of-change of both $A_{\perp}$ and B, that is, on the rate of charge of their product. This should, bring to mind the notion of the flux of the field—the product of field and area where the penetration is perpendicular. Accordingly, the flux of the magnetic field through the wire loop is

$$\Phi_m = B_{\perp} A = BA_{\perp} = BA\cos\theta$$

More generally, if varies in space as it's likely to, the flux of the magnetic field through any open area A bounded by the conducting loop (Figure 4-4) is given by

$$\Phi_m = \iint_A B \, dS \qquad (4.1)$$

The induced emf, developed around the loop, is then

$$emf = -\frac{d\Phi_m}{dt} \qquad (4.2)$$

We should not, however, get too involved with the image of wires and current and emf. Our present concern is with the electric and magnetic fields themselves.

Figure 4-4 B-field through an open area A

In very general terms, an emf is a potential difference, and that's a potential-energy difference per unit charge. A potential-energy difference per unit charge corresponds to work done per unit charge, which is force per unit charge times distance, and that's electric field times distance. The emf exists only as a result of the presence of an electric field

$$emf = \oint_C E \, dl \qquad (4.3)$$

taken around the closed curve C, corresponding to the loop. Equating Eq. (4.2) and Eq. (4.3), and making use of Eq. (4.1), we get

$$\oint_C E \, dl = -\frac{d}{dt}\iint_A B \, dS \qquad (4.4)$$

We began this discussion by examining a conducting loop, and have arrived at Eq. (4.4); this expression, except for the path C, contains no reference to the physical loop. In fact, the path can be chosen arbitrarily and need not be within, or anywhere near, a conductor. The electric field in Eq. (4.4) arises not from the presence of electric charges but rather from the time-varying magnetic field. With no charges to act as sources or sinks, the field lines close on themselves, forming loops (Figure 4-5).

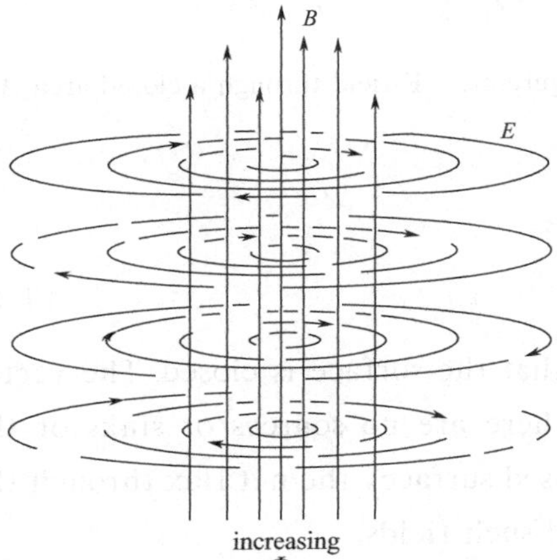

Figure 4-5 A time-varying B-field Surrounding each point where Φ_m is changing, the E-field forms closed loops.

We are interested in EM waves traveling in space where are no wire loops, and the magnetic flux changes because B changes. The Induction Law [Eq. (4.4)] can then be rewritten as

$$\oint_C E dl = \iint_A \frac{\partial B}{\partial t} dS \qquad (4.5)$$

A partial derivative with respect to t is taken because B is usually a function of the space variables. This expression in itself is rather fascinating, since it indicates that a time-varying magnetic field will have an electric field associated with it.

4.3.2 Gauss's Law-Electric

Another fundamental law of electromagnetism is named after the German mathematician Karl Friedrich Gauss (1777—1855). Gauss's Law is about the relationship between the flux of the electric field and the sources of that flux, charge. The ideas derive from fluid dynamic, where both the concepts of field and flux were introduced. The flow of a fluid, as represented by its velocity field, is depicted via streamlines, much as the electric field is pictured via field lines. Figure 4-6 shows a portion of a moving fluid within which there is a region isolated by an imaginary closed surface.

The discharge rate , or volume flux (A_{υ}), is the volume of fluid flowing past a point in the tube per unit time. The volume flux through both end surfaces is equal in magnitude—what flows in per second flows out per second. The net fluid flux (into and out of the closed area) summed over all the surfaces equals zero. If, however, a small pipe is inserted into the region either sucking out fluid (a sink) or delivering fluid (a source), the net flux will then be nonzero.

To apply these ideas to the electric field, consider an imaginary closed area. A placed in some arbitrary electric field, as depicted in Figure 4-7.

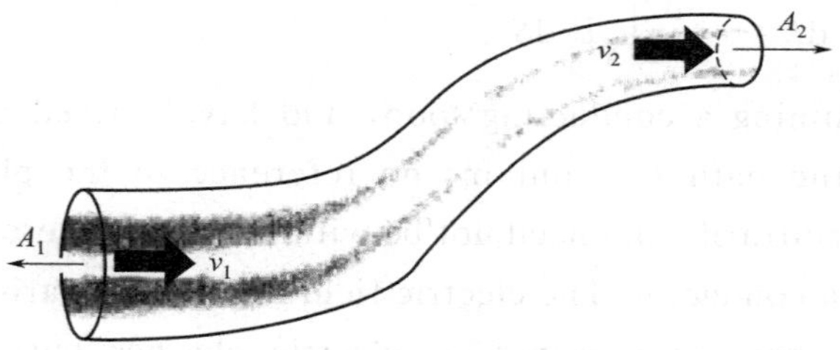

Figure 4-6 A tube of fluid flow

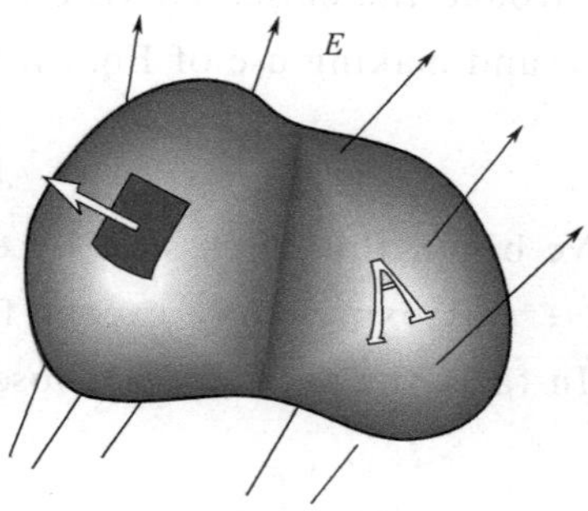

Figure 4-7 E-field through a closed area A

Notice how the area vectors on the E-fields point outward.

The flux of electric field through A is taken to be

$$\Phi_E = \oint\!\!\oint_A E\,\mathrm{d}S \tag{4.6}$$

The circled double integral serves as a reminder that the surface is closed. The vector $\mathrm{d}S$ is in the direction of an outward normal. When there are no sources or sinks of the electric field within the region encompassed by the closed surface, the net flux through the surface equals zero—that much is a general rule for all such fields.

In order to find out what would happen in the presence of internal sources and sinks, consider a spherical surface of radius r centered on and surrounding a positive point-charge (q) in vacuum. The E-field is everywhere outwardly radial, and at any distance r it is entirely perpendicular to the surface: $E = E_{\perp}$ and so $\Phi_E = \oint\!\!\oint_A E_{\perp}\,\mathrm{d}S \oint\!\!\oint_A E\,\mathrm{d}S$

Moreover, since E is constant over the surface of the sphere, it can be taken

$$\Phi_E = E\oint\!\!\oint_A dS = E4\pi r^2$$

But we know from Coulomb's Law that the point-charge has an electric field given by

$$E = \frac{1}{4\pi\varepsilon_0}\frac{q}{r^2}$$

and so

$$\Phi_E = \frac{q}{\varepsilon_0}$$

This is the electric flux associated with a single point-charge q within the closed surface. Since all charge distributions are made up of point-charges, it's reasonable that the net flux due to a number of charges contained within any closed area is

$$\Phi_E = \frac{1}{\varepsilon_0}\sum q$$

Combining the two equations for Φ_E, we get Gauss's Law

$$\oint\!\!\oint_A E dS = \frac{1}{\varepsilon_0}\sum q$$

In order to apply the calculus , it's useful to approximate the charge distribution as being continuous. Then if the volume enclosed by A is V and the charge distribution has a density ρ, Gauss's Law becomes

$$\oint\!\!\oint_A E dS = \frac{1}{\varepsilon_0}\iiint_V \rho dV \tag{4.7}$$

Electric Permittivity

For the special case of vacuum, the electric permittivity of free space is given by $\varepsilon_0 = 8.8542\times10^{-12} C^2/(N\cdot m^2)$. If the charge is embedded in some material medium, its permittivity (ε) will appear in Eq. (4.7) instead of ε_0. One function of the permittivity in Eq. (4.7) is, of course, to balance out the units. There it's the medium-dependent proportionality constant between the device's capacitance and its geometric characteristics. Indeed ε is often measured by a procedure in which the material under study is placed within a capacitor. Conceptually, the permittivity embodies the electrical behavior of the medium: in a sense, it is a measure of the degree to which the material is permeated by the electric field in which it is immersed.

In the early days of the development of the subject, people in various areas worked in different systems of units, a state of affairs leading to some obvious difficulties. This necessitated the tabulation of numerical values for ε in each of the different systems, which was, at best, a waste of time. The same problem regarding densities was neatly avoided by using specific gravity. Thus it was advantageous to tabulate values not of ε but of a new related quantity that is independent of the system of units being used. Accordingly, we define that is K_E as $\varepsilon/\varepsilon_0$. This is the dielectric constant, and it is appropriately unitless. The permittivity of a material can be expressed in terms of ε_0 as

$$\varepsilon = K_E \varepsilon_0 \tag{4.8}$$

Our interest in K_E anticipates the fact that the permittivity is related to the speed of light in dielectric materials, such as glass, air, and quartz.

4.3.3 Gauss's Law-Magnetic

There is no known magnetic counterpart to the electric charge; that is, no isolated magnetic poles have ever been found, despite extensive searching, even in lunar soil samples. Unlike the electric field, the magnetic field B does not diverge from or converge toward some kind of magnetic charge (a monopole source or sink). Magnetic fields can be described in terms of current distributions. Indeed, we might envision an elementary magnet as a small current loop in which the lines of B are continuous and closed. Any closed surface in a region of magnetic field would accordingly have an equal number of lines of B entering and emerging from it.

This situation arises from the absence of any monopoles within the enclosed volume. The flux of magnetic field Φ_m through such a surface is zero, and we have the magnetic equivalent of Gauss's Law

$$\Phi_m = \oiint_A B\, dS = 0 \tag{4.9}$$

4.3.4 Ampere's Circuital Law

Another equation that will be of great interest is associated with Andre Marie Ampere (1775—1836). Known as the Circuital Law, its physical origins are a little obscure—it will take a bit of doing to justify it, but it's worth it. Accordingly, imagine a straight current—carrying wire in vacuum and the circular B-field surrounding it.

We know from experiments that the magnetic field of a straight wire carrying a current i is $B = \mu_i / 2\pi r$. Now, suppose we put ourselves back in time to the nineteenth century when it was common to think of magnetic charge (q_m). Let's define this monopole charge so that it experiences a force when placed in a magnetic field B equal to $q_m B$ in the direction of B , just as an electric charge q_e experiences a force $q_e E$. Suppose we carry this north-seeking monopole around a closed circular path perpendicular to and centered on a current-carrying wire and determine the work done in the process. Since the direction of the force changes, because B changes direction, we will have to divide the circular path into tiny segments (Δ_l) and sum up the work done over each. Work is the component of the force parallel to the displacement times the displacement: $\Delta W = q_m B_{//} \Delta_l$, and the total work done by the field is $\sum q_m B_{//} \Delta_L$. In this case, B is everywhere tangent to the path, so that $B_{//} = B = \mu_i / 2\pi r$, which is constant around the circle. With both q_m and B constant, the summation becomes

$$\sum q_m B_{//} \Delta_l = q_m \sum B \Delta_l = q_m B 2\pi r$$

where $\Sigma\Delta_1 = 2\pi r$ is the circumference of the circular path.

If we substitute for B the equivalent current expression, which varies inversely with r, the radius cancels—the work is independent of the circular path taken. Since no work is done in traveling perpendicular to B, the work must be the same if we move q_m (out away from the wire or in toward it) along a radius, carrying it from one circular segment to another as we go around. Indeed, W is independent of path altogether—the work will be the same for any closed path encompassing the current. Putting in the current expression for B and canceling the "charge" q_m, we get the rather remarkable expression

$$\Sigma B_{//}\Delta_1 = \mu_0 i$$

which is to be summed over any closed path surrounding i. The magnetic charge has disappeared, which is nice, since we no longer expect to be able to perform this little thought experiment with a monopole. Still, the physics was consistent, and the equation should hold, monopoles or no. Moreover, if there are several current-carrying wires encompassed by the closed path, their fields will superimpose and add, yielding a net field. The equation is true for the separate fields and must be true as well for the field. Hence

$$\Sigma B_{//}\Delta_1 = \mu_0 \Sigma i$$

As $\Delta_1 \to 0$, the sum becomes an integral around a closed path

$$\oint_C B\mathrm{d}l = \mu_0 \Sigma i$$

Today this equation is known as Ampere's Law, though at one time it was commonly referred to as the "work rule". It relates a line integral of B tangent to a closed curve C, with the total current i passing within the confines of C.

When the current has a nonuniform cross section, Ampere's Law is written in terms of the current density or current per unit area J, integrated over the area

$$\oint_C B\mathrm{d}l = \mu_0 \iint_A J\mathrm{d}S \tag{4.10}$$

The open surface A is bounded by C (Figure 4-8). The quantity μ_0 is called the permeability of free space and it's defined as $4\pi\times10^{-7}$ $\mathrm{N\cdot S^2/C^2}$. When the current is embedded in a material medium, its permeability (μ) will appear in Eq. (4.10). As in Eq. (4.8)

$$\mu = K_m \mu_0 \tag{4.11}$$

with K_m being the dimensionless relative permeability.

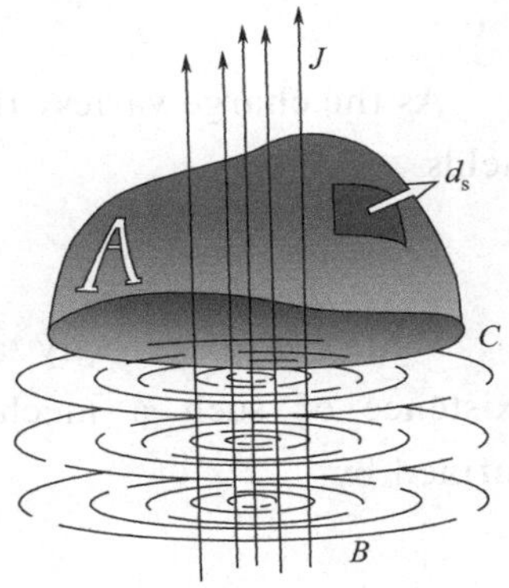

Figure 4-8 Current density through an open area A

Eq. (4.10), though often adequate, is not the whole truth. Ampere's Law is not particular about the area used, provided it's bounded by the curve C, which makes for an obvious problem when charging a capacitor, as shown in Figure 4-9 (a). If flat area A_1 is used, a net current of i flows through it and there is a B-field along curve C. The right side of Eq. (4.10) is nonzero, so the left side is nonzero. But if area A_2 is used instead to encompass C, no net current passes through it and the field must now be zero, even though nothing physical has actually changed. Something is obviously wrong!

Moving charges are not the only source of a magnetic field. While charging or discharging a capacitor, one can measure a B-field in the region between its plates [Figure 4-9 (b)], which is indistinguishable from the field surrounding the leads , even though no electric current actually traverses the capacitor. Notice, however, that if A is the area of each plate and Q is the charge on it

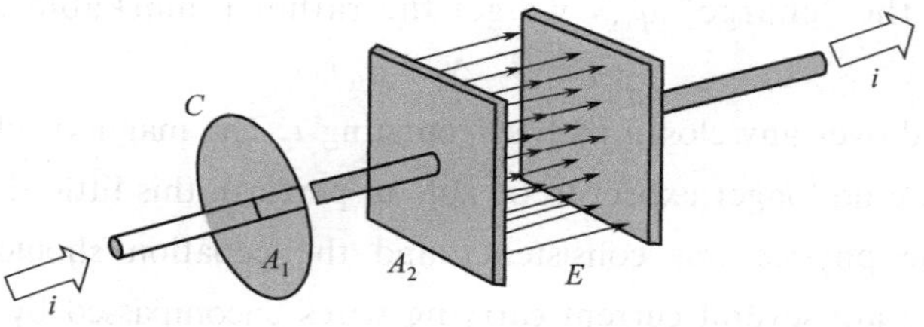

(a) Ampere's Law is indifferent to which area A_1 or A_2 is bounded by the path C. Yet a current passes through A_1,and not through A_2,and that means something is very wrong

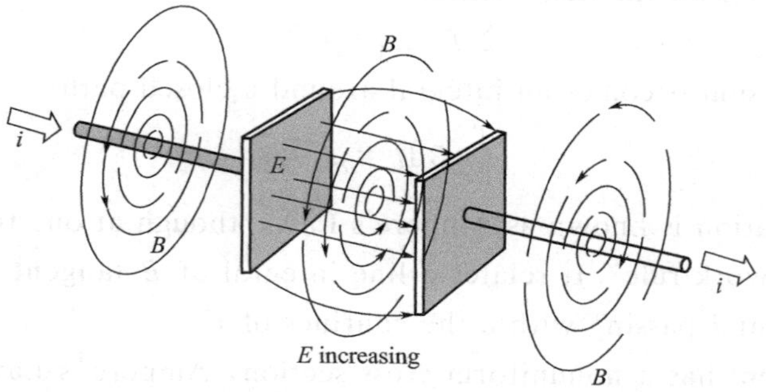

(b) B-field concomitant with a time-varying E-field in the gap of a capacitor

Figure 4-9 The change of Ampere's Law when a capacitor exists

$$E = \frac{Q}{\varepsilon A}$$

As the charge varies, the electric field changes, and taking the derivative of both sides yields

$$\varepsilon \frac{\partial E}{\partial t} = \frac{i}{A}$$

Which is effectively a E current density. James Clerk Maxwell hypothesized the existence of such a mechanism, which he called the displacement current density, defined by

$$JD \equiv \varepsilon \frac{\partial E}{\partial t} \tag{4.12}$$

The restatement of Ampere's Law as

$$\oint_C B\mathrm{d}l = \mu \iint_A \left(J + \varepsilon \frac{\partial E}{\partial t}\right)\mathrm{d}S \tag{4.13}$$

was one of Maxwell's greatest contributions. It points out that even when $J = 0$, a time-varying E-field will be accompanied by a B-field (Figure 4-7).

4.3.5 Maxwell's Equations

The set of integral expressions given by Eq. (4.5), Eq. (4.7), Eq. (4.9), and Eq. (4.13) have come to be known as Maxwell's Equations. Remember that these are the generalizations of experimental results. The simplest statement of Maxwell's Equations applies to the behavior of the electric and magnetic fields in free space, where $E=\varepsilon_0$, $\mu=\mu_0$ and both ρ and J are zero. In that instance

$$\oint_{C} E\mathrm{d}l = -\iint_{A} \frac{\partial B}{\partial t}\mathrm{d}S \tag{4.14}$$

$$\oint_{C} B\mathrm{d}l = \mu_0\varepsilon_0\iint_{A} \frac{\partial E}{\partial t}\mathrm{d}S \tag{4.15}$$

$$\oiint_{A} B\,\mathrm{d}S = 0 \tag{4.16}$$

$$\oiint_{A} E\,\mathrm{d}S = 0 \tag{4.17}$$

Observe that except for a multiplicative scalar, the electric and magnetic fields appear in the equations with a remarkable symmetry. However E affects B, B will in turn affect E. The mathematical symmetry implies a good deal of physical symmetry. The most significant outcome of this theory is the prediction of the existence of EM waves.

4.4 The Properties of Electromagnetic Waves

The physics of EM radiation is electrodynamics, a subfield of electromagnetism. EM radiation exhibits both wave properties and particle properties at the same time (wave-particle duality). The wave characteristics are more apparent when EM radiation is measured over relatively large timescales and over large distances, and the particle characteristics are more evident when measuring small distances and timescales. Both characteristics have been confirmed in a large number of experiments.

There are experiments in which the wave and particle natures of EM waves appear in the same experiment, such as the diffraction of a single photon. When a single photon is sent through two slits, it passes through both of them interfering with itself, as waves do, yet is detected by a photomultiplier or other sensitive detector only once. Similar selfinterference is observed when a single photon is sent into a Michelson interferometer or other interferometers.

4.4.1 Wave Model

An important aspect of the nature of light is frequency. The frequency of a wave is its rate of oscillation and is measured in hertz, the SI unit of frequency, equal to one oscillation per second. Light usually has a spectrum of frequencies which sum together to form the resultant wave. Different frequencies undergo different angles of refraction.

A wave consists of successive troughs and crests, and the distance between two adjacent crests or troughs is called the wavelength. Waves of the EM spectrum vary in size, from very long radio waves (the size of buildings) to very short gamma rays (smaller than atom nuclei). Frequency is inversely proportional to wavelength, according to the equation

$$v = f\lambda \tag{4.18}$$

where v is the speed of the wave (c in vacuum, or less in other media), f is the frequency and λ is the wavelength. As waves cross boundaries between different media, their speeds change but their frequencies remain constant.

Interference is the superposition of two or more waves resulting in a new wave pattern. If the fields have components in the same direction, they constructively interfere, while opposite directions cause destructive interference. The energy in EM waves is sometimes called radiant energy.

4.4.2 Particle Model

Because energy of an EM wave is quantized, in the particle model of EM radiation, a wave consists of discrete packets of energy, or quanta, called photons. The frequency of the wave is proportional to the magnitude of the particle's energy. Moreover, because photons are emitted and absorbed by charged particles, they act as transporters of energy. The energy per photon can be calculated by Planck's equation

$$E = hf \tag{4.19}$$

where E is the energy, h is Planck's constant.

As a photon is absorbed by an atom, it excites an electron, elevating it to a higher energy level. If the energy is great enough, so that the electron jumps to a high enough energy level, it may escape the positive pull of the nucleus and be liberated from the atom in a process called photoionization. Conversely, an electron that descends to a lower energy level in an atom emits a photon of light equal to the energy difference. Since the energy levels of electrons in atoms are discrete, each element emits and absorbs its own characteristic frequencies.

Together , these effects explain the absorption spectra of light. The dark bands in the spectrum are due to the atoms in the intervening medium absorbing different frequencies of the light. The composition of the medium through which the light travels determines the nature of the absorption spectrum. For instance, dark bands in the light emitted by a distant star are due to the atoms in the star' s atmosphere. These bands correspond to the allowed energy levels in the atoms. A similar phenomenon occurs for emission. As the electrons descend to lower energy levels, a spectrum is emitted that represents the jumps between the energy levels of the electrons. This is manifested in the emission spectrum of nebulae. Today, scientists use this phenomenon to observe what elements a certain star is composed of. It is also used in the determination of the distance of a star, using the socalled red shift.

4.4.3 Speed of Propagation

Any electric charge which accelerates, or any changing magnetic field, produces EM radiation. EM information about the charge travels at the speed of light. Accurate treatment thus incorporates a concept known as retarded time (as opposed to advanced time, which is unphysical in light of causality), which adds to the expressions for the electrodynamic electric field and magnetic field. These extra terms are responsible for EM radiation. When any wire (or other conducting object such as an antenna) conducts alternating current, EM radiation is propagated at the same frequency as the electric current. Depending on the circumstances, it may behave as a wave or as particles. As a wave, it is characterized by a velocity (the speed of light), wavelength, and frequency.

One rule is always obeyed regardless of the circumstances: EM radiation in a vacuum always travels at the speed of light, relative to the observer, regardless of the observer's velocity. (This observation led to Albert Einstein's development of the theory of special relativity.)

In a medium (other than vacuum), velocity of propagation or refractive index are considered, depending on frequency and application. Both of these are ratios of the speed in a medium to speed in a vacuum.

4.5 Electromagnetic Spectra and Applications

EM energy at a particular wavelength λ (in vacuum) has an associated frequency f and photon energy E. Thus, the EM spectrum may be expressed equally well in terms of any of these three quantities. They are related according to the equations

$$E = hc/\lambda$$

So, high-frequency EM waves have a short wavelength and high energy; low-frequency waves have a long wavelength and low energy.

The EM spectrum is the range of all possible EM radiation. The "EM spectrum" (usually just spectrum) of an object is the frequency range of EM radiation with wavelengths from thousands of kilometers down to fractions of the size of an atom. Generally, EM radiation is classified by wavelength into electrical energy, radio, microwave, terahertz, infrared, the visible region we perceive as light, ultraviolet, X-rays and gamma rays. It is commonly said that EM waves beyond these limits are uncommon, although this is not actually true. The short wavelength limit is likely to be the Planck length, and the long wavelength limit is the size of the universe itself, though in principle the spectrum is infinite.

4.5.1 Radio waves

Radio waves generally are utilized by antennas of appropriate size (according to the principle of resonance), with wavelengths ranging from hundreds of meters to about one millimeter. They are used for transmission of data, via modulation. Television, mobile phones, wireless networking and amateur radio all use radio waves. The use of the radio spectrum is regulated by many governments through frequency allocation.

Radio waves can be made to carry information by varying a combination of the amplitude, frequency and phase of the wave within a frequency band. When EM radiation impinges upon a conductor, it couples to the conductor, travels along it, and induces an electric current on the surface of that conductor by exciting the electrons of the conducting material. This effect (the skin effect) is used in antennas.

4.5.2 Microwaves

The super high frequency (SHF) and extremely high frequency (EHF) of microwaves come next up the frequency scale. Microwaves are waves which are typically short enough to employ tubular metal waveguides of reasonable diameter. Microwave energy is produced with klystron and magnetron tubes, and with solid state diodes such as Gunn and IMPATT devices. Microwaves are absorbed by molecules that have a dipole moment in liquids. In a microwave oven, this effect is used to heat food. Low-intensity microwave radiation is used in Wi-Fi, although this is at intensity levels unable to cause thermal heating.

Volumetric heating, as used by microwaves, transfer energy through the material electro-magnetically, not as a thermal heat flux. The benefit of this is a more uniform heating and reduced heating time; microwaves can heat material in less than 1% of the time of conventional heating methods.

When active, the average microwave oven is powerful enough to cause interference at close range with poorly shielded EM fields such as those found in mobile medical devices and cheap consumer electronics.

4.5.3 Terahertz radiation

Terahertz radiation is a region of the spectrum between far infrared and microwaves. Until recently, the range was rarely studied and few sources existed for microwave energy at the high end of the band (sub-millimeter waves or so-called terahertz waves), but applications such as imaging and communications are now appearing. Scientists are also looking to apply terahertz technology in the armed forces, where high frequency waves might be directed at enemy troops to incapacitate their electronic equipment.

4.5.4 Infrared radiation

The infrared part of the EM spectrum covers the range from roughly 300 GHz (1 mm) to 400 THz (750 nm). It can be divided into three parts:

Far-infrared, from 300 GHz (1 mm) to 30 THz (10 μm). The lower part of this range may also be called microwaves. This radiation is typically absorbed by so-called rotational modes in gas-phase molecules, by molecular motions in liquids, and by phonons in solids. The water in the earths' atmosphere absorbs so strongly in this range that it renders the atmosphere effectively opaque. However, there are certain wavelength ranges ("windows") within the opaque range which allow partial transmission, and can be used for astronomy. The wavelength range from approximately 200 μm up to a few mm is often referred to as "sub-millimeter" in astronomy, reserving far infrared for wavelengths below 200 μm.

Mid-infrared, from 30 to 120 THz (10 to 2.5 μm), hot objects (black-body radiators) can radiate strongly in this range. It is absorbed by molecular vibrations, where the different atoms in a molecule vibrate around their equilibrium positions. This range is sometimes called the fingerprint region since the mid-infrared absorption spectrum of a compound is very specific for that compound.

Near-infrared, from 120 to 400 THz (2500 to 750 nm), physical processes that are relevant for this range are similar to those for visible light.

4.5.5 Visible radiation (light)

Above infrared in frequency comes visible light. This is the range in which the sun and stars similar to it emit most of their radiation. It is probably not a coincidence that the human eye is sensitive to the wavelengths that the sun emits most strongly. Visible light (and near-infrared light) is typically absorbed and emitted by electrons in molecules and atoms that move from one energy level to another. The light we see with our eyes is really a very small portion of the EM spectrum. A rainbow shows the optical (visible) part of the EM spectrum; infrared (if you could see it) would be located just beyond the red side of the rainbow with ultraviolet appearing just beyond the violet end.

EM radiation with a wavelength between 380 nm and 760 nm (790～400 THz) is detected by the human eye and perceived as visible light. Other wavelengths, especially near infrared (longer than 760 nm) and ultraviolet (shorter than 380 nm) are also sometimes referred to as light, especially when the visibility to humans is not relevant.

If radiation having a frequency in the visible region of the EM spectrum reflects off of an object, say, a bowl of fruit, and then strikes our eyes, this results in our visual perception of the scene. Our brain's visual system processes the multitude of reflected frequencies into different shades and hues, and through this not-entirely-understood psychophysical phenomenon, most people perceive a bowl of fruit.

At most wavelengths, however, the information carried by EM radiation is not directly detected by human senses. Natural sources produce EM radiation across the spectrum, and our technology can also manipulate a broad range of wavelengths. Optical fiber transmits light which, although not suitable for direct viewing, can carry data that can be translated into sound or an image. The coding used in such data is similar to that used with radio waves.

4.5.6 Ultraviolet light

Next in frequency comes ultraviolet (UV). This is radiation whose wavelength is shorter than the violet end of the visible spectrum, and longer than that of an X-ray.

Being very energetic, UV can break chemical bonds, making molecules unusually reactive or ionizing them (see photoelectric effect), in general changing their mutual behavior. Sunburn, for example, is caused by the disruptive effects of UV radiation on skin cells, which is the main cause of skin cancer, if the radiation irreparably damages the complex DNA molecules in the cells (UV radiation is a proven mutagen). The sun emits a large amount of UV radiation, which could quickly turn earth into a barren desert. However, most of it is absorbed by the atmosphere's ozone layer before reaching the surface.

4.5.7 X-rays

After UV come X-rays, which are also ionizing, but due to their higher energies they can also interact with matter by means of the Compton effect. Hard X-rays have shorter wavelengths than soft X-rays. As they can pass through most substances, X-rays can be used to "see through" objects, most notably diagnostic X-ray images in medicine (a process known as radiography), as well as for high-energy physics and astronomy. Neutron stars and accretion disks around black holes emit X-rays, which enable us to study them. X-rays are given off by stars, and strongly by some types of nebulae.

4.5.8 Gamma rays

After hard X-rays come gamma rays, which were discovered by Paul Villard in 1900. These are the most energetic photons, having no defined lower limit to their wavelength. They are useful to astronomers in the study of high energy objects or regions, and find a use with physicists thanks to their penetrative ability and their production from radioisotopes. Gamma rays are also used for the irradiation of food and seed for sterilization, and in medicine they are used in radiation cancer therapy and some kinds of diagnostic imaging such as PET scans. The wavelength of gamma rays can be measured with high accuracy by means of Compton scattering.

Note that there are no precisely defined boundaries between the bands of the EM spectrum. Radiation of some types have a mixture of the properties of those in two regions of the spectrum. For example, red light resembles infrared radiation in that it can resonate some chemical bonds.

New Words and Expressions

atom smashers 核粒子加速器

bewildering 令人困惑的，使人混乱的

calculus 微积分

celestial mechanics 天体力学

charge 电荷
electromotive force 电动势
electrostatic 静电的，静电学的
ferromagnetism 铁磁性
formulate 用公式表示
galvanometer 检流计
induction 感应，感应现象
Michelson interferometer 迈克尔逊干涉仪
molecule 分子
nebula 星云（nebulae 是其复数形式）
orthogonal 直角的，直交的
oscillate 振荡
photon 光子
quantitative study 定量研究
radio astronomy 电波（无线电）天文学
reciprocal effect 电磁感应
reciprocal 相互的，交互的，相反的
speculate 深思，推测
spherical 球的，球面的，球状的
sterilization 消毒，灭菌
streamline 流线型的
superposition 重叠，叠合，重合

Translation

1. If the magnitude and the location of the electric charge do not change with time, the electric field produced by the charge will also be constant over time.

翻译：如果电荷的大小和位置不随时间变化，电荷产生的电场也将随时间而保持恒定。

2. That is, the electric and magnetic fields are orthogonal (perpendicular) to each other, and they are orthogonal to the direction of propagation of the photon.

翻译：也就是说，电场和磁场相互正交（垂直），并且它们与光子的传播方向正交。

3. From a classical perspective, the EM field can be regarded as a smooth, continuous field, propagated in a wavelike manner; whereas, from a quantum mechanical perspective, the field is seen as quantized, being composed of individual particles.

翻译：从经典的角度看，电磁场可以看作是一个平滑的、连续的场，以波的方式传播；而从量子力学的角度看，场被看作是量子化的，由单个粒子组成。

4. Of course there are other phenomena that we recognize today to be of EM origin, which have been observed since the beginning of time.

翻译：当然，我们今天认识到的其他现象是电磁导致的，这些现象从一开始就被观察到了。

5. In 1787 Coulomb proposed a law of force for charges that, like Sir Isaac Newton's law of gravitation, varied inversely as the square of the distance.

翻译：1787 年，库仑提出了电荷间力的定律，它和艾萨克·牛顿的万有引力定律一样，该力与距离的平方成反比。

6. At the beginning of the nineteenth century, the electrochemical cell was invented by Alessandro Volta, professor of natural philosophy at the University of Pavia in Italy.

翻译：19 世纪初，电化学电池由意大利帕维亚大学自然哲学教授亚历山德罗·伏打发明。

7. In 1820 at the University of Copenhagen, Hans Christian Oersted, professor of physics, made the momentous discovery that an electric current in a wire could deflect a magnetic needle.

翻译：1820 年，哥本哈根大学物理学教授汉斯·克里斯蒂安·奥斯特有了一个重大发

现：导线中的电流可以使磁针偏转。

8. The favoured explanation, which was provided by Ampere, was that central forces were the cause.

翻译：安培提供的最有利的解释是，中心力是原因。

9. Subsequently, in 1831, the British scientist Michael Faraday demonstrated the reciprocal effect, in which a moving magnet in the vicinity of a coil of wire produced an electric current.

翻译：随后，1831 年英国科学家迈克尔·法拉第演示了这种相互作用，即在一个线圈附近移动磁铁会产生电流。

10. As is typical of abstract mathematical reasoning, Maxwell saw in his equations a certain symmetry that suggested the need for an additional term, involving the time rate of change of the electric field.

翻译：作为典型的抽象数学推理，麦克斯韦在他的方程中看到了某种对称性，这意味着需要一个附加变量，来概括电场的时间变化率。

11. Using the values of known experimental constants obtained solely from measurements of charges and currents, Maxwell deduced that the speed of propagation was equal to speed of light.

翻译：麦克斯韦利用仅从电荷和电流测量中获得的已知实验常数值，推导出传播速度等于光速。

12. Maxwell's theory was not accepted by scientists immediately, in part because it had been derived from a bewildering collection of mechanical analogies and difficult mathematical concepts.

翻译：麦克斯韦的理论并没有立即被科学家们所接受，部分原因是它来源于一系列令人困惑且机械性的类比和复杂的数学概念。

Unit 5 Lasers

5.1 Amplification of Light

Suppose we locate an amplifying rod (or tube of gas or liquid) between two mirrors. One mirror is partially transparent; both are aligned parallel to one another and perpendicular to the axis of the rod. The optical length of the cavity thus formed is d, the reflectance of the partially transmitting mirror is R, and the (intensity) gain of the rod is G.

Initially, the only light emitted is that arising from fluorescence or spontaneous emission. As we have noted, the fluorescent emission is not directional, but some of this light will travel along the cavity's axis. The following heuristic argument shows how amplified fluorescence brings about laser emission.

Consider a wave packet that is emitted along the axis by a single atom. The packet undergoes many reflections from the mirrors. After each round trip in the cavity, it is amplified by G^2 and diminished by R. If the net round-trip gain exceeds 1

$$G^2 R > 1 \tag{5.1}$$

then the wave grows almost without limit. Only waves that travel parallel to the axis experience such continuous growth; the result is therefore a powerful, directional beam. The useful output of the laser is the fraction that escapes through the partial reflector or output mirror.

There is a second condition necessary for lasing. Consider again the wave packet emitted along the axis. Its coherence length is great compared with the optical length of the cavity. In a sense, the atom therefore emits the packet over a finite time. Because of multiple reflections, the packet returns many times to the atom before the emission is completed. If the packet returns out of phase with the wave that is still being emitted by the atom, it will interfere destructively with that wave and effectively terminate the emission. We can, if we wish, say that such a wave has been reabsorbed, but the effect is as if the wave never existed.

The only waves that exist are, therefore, those for which constructive interference occurs

$$m\lambda = 2d \tag{5.2}$$

As a result of the large number of reflections, only wavelengths quite close to $2d/m$ exist. This is analogous to the sharpness of multiple-beam interference fringes. In general, there are many such wavelengths within the fluorescent line-width of the source, so the value of d is not at all critical.

It is convenient now to consider the cavity separately from the amplifying medium. We may later combine their properties to account for the properties of the emitted light. In the following sections, we shall treat the properties of the amplifier and the cavity in somewhat greater detail.

5.2 Optical Resonators

An optical resonator consists of two mirrors facing each other, as in the Fabry-Perot interferometer. Both mirrors have highly reflecting coatings. Fabry-Perot interferometers have narrow transmissions pass bands at discrete optical frequencies.

5.2.1 Longitudinal Modes

To begin, consider a plane wave that originates inside a Fabry-Perot resonator, with amplitude A_0. After a large number of reflections have occurred, the total field inside the cavity is

$$E = A_0(1 + r^2 e^{-i\Phi} r^4 e^{-2i\Phi} + \cdots) \tag{5.3}$$

for the wave traveling to the right. r is the amplitude reflection coefficient of both mirrors, and Φ ($=2kd$) is the phase change associated with one round trip. The wave is assumed to travel normal to the mirrors.

We calculate the sum of the geometric series and find that the intensity I inside the cavity is

$$\frac{I_0}{1 + F\sin^2 \dfrac{\Phi}{2}} \tag{5.4}$$

where $F = 4R/(1-R)^2$ and R is the reflectance. This is precisely the expression for the transmittance of a Fabry-Perot interferometer. I has value I_0 when the familiar condition for constructive interference satisfies, i. e.,

$$q\lambda = 2d \tag{5.5}$$

where q is an integer.

Eq. (5.5) corresponds to a spacing of the two mirrors of q half-wavelengths. Each value of q corresponds to a different longitudinal mode of the interferometer having a frequency of

$$\nu = q\frac{c}{2d} \tag{5.6}$$

The frequency separation between modes follows from

$$\Delta\nu = \frac{c}{2d}(q+1) - \frac{c}{2d}q = \frac{c}{2d} \tag{5.7}$$

The frequency interval $\Delta\nu$ is called the free spectral range of the interferometer, or in terms of wavelength

$$\Delta\lambda = \frac{\lambda^2}{2d} \tag{5.8}$$

The longitudinal modes form an equally spaced comb of resonant frequencies, as shown in Figure 5-1.

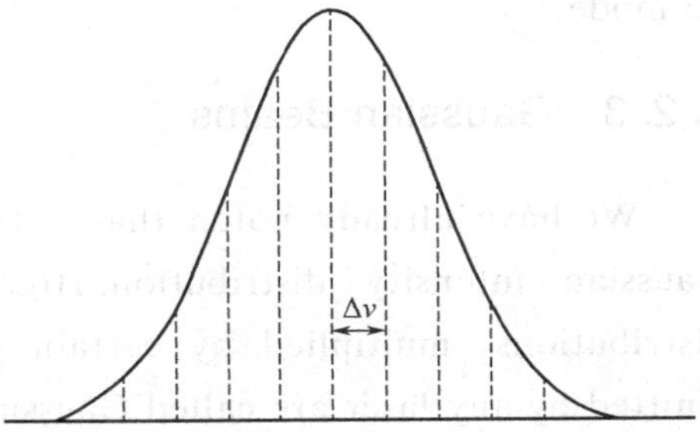

Figure 5-1 Longitudinal modes in a stable laser

5. 2. 2 Transverse Modes

These are most easily understood in terms of a cavity such as a confocal cavity. The confocal cavity has two identical mirrors with a common focus at *F*. Roughly speaking, we can define a transverse mode as the electric field distribution that is associated with any geometrical ray that follows a closed path (We will not consider cavities that do not allow a ray to follow a closed path; these are known as unstable resonators). Naturally, the ray will not describe the precise field distribution because of the effects of diffraction.

The simplest mode in a confocal cavity is described by a ray that travels back and forth along the axis. This is the 00 mode. Because of diffraction, the actual intensity distribution is outlined. The output of a laser oscillating in this mode is a spherical wave with a Gaussian intensity distribution. The beam width is usually expressed as the radius w_0 at which the beam intensity falls to $1/e^2$ of its maximum value.

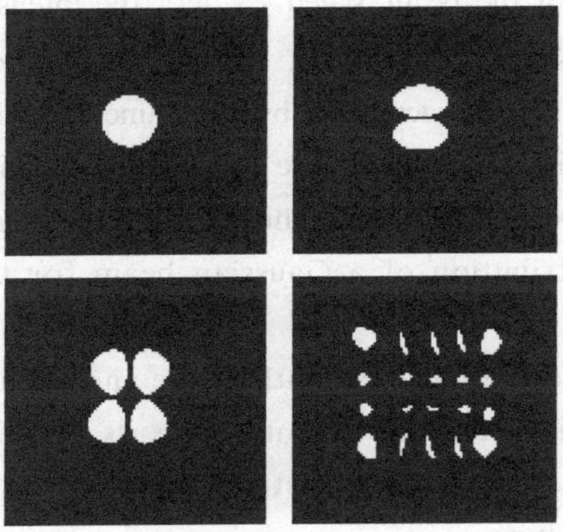

Figure 5-2 Laser transverse-mode patterns

The next-simplest mode is also shown in Figure 5-2. This mode will oscillate only if the aperture (which is often defined by the laser tube) is large enough. Higher-order modes correspond to closed paths with yet-higher numbers of reflections required to complete a round trip. In Figure 5-2, upper left corresponds to 00 mode; upper right represents 01 mode; lower left denotes 11 mode; lower right corresponds to coherent superposition of two or more transverse modes.

Transverse-mode patterns are labeled according to the number of minima that are encountered when the beam is scanned horizontally (first number) and then vertically (second number). The modes shown all have rectangular symmetry; such rectangular modes characterize nearly all lasers, including those with cylindrical rods or tubes. In practice, the higher-order modes have greater loss (due to diffraction) than the 00 mode. If a laser oscillates in a certain high-order mode, it also emits all modes with lower order. Such a multimode laser provides considerable power compared with that of a 00 or single-mode laser. Nevertheless, a single-mode laser is often desirable. In many gas lasers, the diameter of the laser tube is chosen small enough that diffraction loss prohibits oscillation of any mode other than the

00 mode.

5.2.3 Gaussian Beams

We have already noted that a laser oscillating in the 00 mode emits a beam with Gaussian intensity distribution. Higher-order modes also exhibit Gaussian intensity distributions, multiplied by certain polynomials (Hermite polynomials). Thus, beams emitted by any laser are called Gaussian beams.

In our earlier treatment of diffraction, we assumed that a uniform wavefront passed through the diffracting screen. We found, for example, that the far-field intensity distribution had an angular divergence of $1.22\lambda/D$ for a circular aperture. That result is not appropriate when Gaussian beams are used because then the intensity distribution is not uniform.

A detailed treatment of propagation of Gaussian beams is left to the references. Here we discuss the general results. The field distribution in any curved-mirror cavity is characterized by a beam waist; in a symmetrical cavity, the beam waist is located in the center of the cavity. The intensity distribution in the plane of the waist is Gaussian for the 00 mode; that is,

$$I(r) = \exp(-2r^2/w_0^2) \tag{5.9}$$

where r is the distance from the center of the beam. For convenience, the intensity is normalized to 1 at the center of the beam. When $r = w_0$, the intensity is $1/e^2$ times the intensity at the center. Higher-order modes are characterized by the same Gaussian intensity distribution, but for the Hermite polynomials mentioned above. Figure 5-3 shows the transverse intensity distribution of a Gaussian beam for the 00 mode.

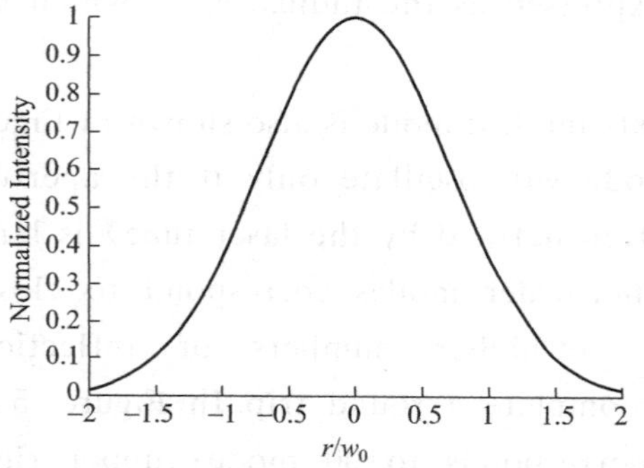

Figure 5-3 Transverse intensity distribution of a fundamental Gaussian beam

The beam propagates, both inside and outside the cavity, in such a way that it retains its Gaussian profile. That is, at a distance z from the beam waist, the intensity distribution is given by the preceding equation with w_0 replaced by $w(z)$, where

$$w(z) = w_0\left[1+\left(\frac{\lambda z}{\pi w_0^2}\right)^2\right]^{1/2} \tag{5.10}$$

At large distances z from the beam waist, the term in parentheses is large compared to 1. In this case, the Gaussian beam diverges with angle θ, where

$$\theta = \frac{\lambda}{\pi w_0} \tag{5.11}$$

This is the far-field divergence of a Gaussian beam. If the beam is brought to a focus with a lens whose diameter is at least $2w_0$, the radius of the focal spot is $\lambda f'/(\pi w_0)$, which is somewhat smaller than the corresponding Airy-disk radius, $0.61\lambda f'/w_0$. In addition, the diffraction pattern is not an Airy disk, but has a Gaussian intensity distribution with no secondary maxima (unless the lens aperture vignettes a significant portion of the incident beam).

The radiation converges toward a beam waist and diverges away from it, as shown in Figure 5-4. Therefore, the wavefront must be planar at the beam waist. At a distance z from the waist, the radius of curvature $R(z)$ of the wavefront is

$$R(z) = z\left[1 + \left(\frac{\pi w_0^2}{\lambda z}\right)\right] \tag{5.12}$$

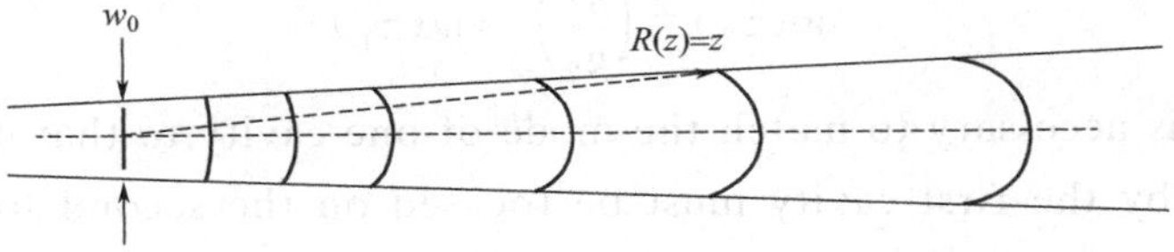

Figure 5-4 Contour of a Gaussian beam near $z = 0$

Only at great distance z from the waist does the beam acquire a radius of curvature equal to z.

Thus far, our comments have been general and apply to any Gaussian beam with a beam waist w_0 located at $z = 0$. To apply the discussion to a laser cavity, we must know the size and position of the beam waist. In a confocal cavity whose mirrors are separated by d, w_0 is given by

$$w_0 = \sqrt{\frac{\lambda d}{2\pi}} \tag{5.13}$$

and the waist ($z = 0$) is located in the center of the cavity, as with any symmetrical, concave-mirror cavity.

When the cavity is not confocal, it is customary to define stability parameters g_1 and g_2 by the equations

$$g_1 = 1 - d/R_1 \tag{5.14}$$

and

$$g_2 = 1 - d/R_2 \tag{5.15}$$

where R_1 and R_2 are the radii of curvature of the mirrors.

To find the size and location of the beam waist, we argue that the radius of curvature of the beam wavefront at the positions of the mirrors must be exactly equal to the radii of the mirrors themselves. If this were not so, then the cavity would not have a stable electric-field distribution; in terms of our earlier ray picture, the rays would not follow a closed path.

We therefore know the radius of curvature of the wavefront at two locations, which we may call z_1 (the distance from the waist to mirror 1) and z_2 (the distance from the waist to mirror 2). Setting $R(z_1) = R_1$ and $R(z_2) = R_2$, we may solve for z_1 and thereby locate the waist. The result is

$$z_1 = [g_2(1 - g_1)/(g_1 + g_2 - 2g_1 g_2)]d \tag{5.16}$$

in terms of the stability parameters.

Similarly, we may solve for the beam waist w_0, which is, in general,

$$w_0 = \left(\frac{\lambda d}{\pi}\right)^{1/2} \left[\frac{g_1 g_2 (1 - g_1 g_2)}{g_1 + g_2 - 2g_1 g_2}\right]^{1/4} \tag{5.17}$$

We may now use the results of Gaussian-beam theory to find the radius $R(z)$ or the spot size $w(z)$ at any location z. In particular, at mirror 1

$$w(z_1) = \left(\frac{\lambda d}{\pi}\right)^{1/2} \left[\frac{g_2}{g_1(1 - g_1 g_2)}\right]^{1/4} \tag{5.18}$$

And at mirror 2

$$w(z_2) = \left(\frac{g_1}{g_2}\right)^{1/2} w(z_1) \tag{5.19}$$

Sometimes it is necessary to match the mode of one cavity to that of another; that is, the mode emitted by the first cavity must be focused on the second so that it becomes a mode of that cavity as well. For example, mode matching may be necessary when a spherical, confocal Fabry-Perot interferometer is used with a laser.

The simplest way to mode match two cavities a and b is to locate the point where their spot sizes $w(z)$ are equal. Then, calculate the beams' radii of curvature $R_a(z)$ and $R_b(z)$ at that point. A lens whose focal length f' is given by

$$1/f' = 1/R_a - 1/R_b \tag{5.20}$$

will match the radii of curvature of the two modes at the point.

Both radius and spot size must be matched to ensure effective mode matching. If both parameters are not matched, for example, power may be lost to higher-order modes in the second cavity. When it is not possible to match two cavities in the simple way described here, it may be necessary to expand or reduce the beam size with a lens before attempting mode matching with a second lens.

5.2.4 Resonator Configurations

The most commonly used laser resonators are composed of two spherical or flat mirrors facing each other, as shown in Figure 5-5. We will first consider the generation of the lowest order mode by such a resonant structure. Once the parameters of the TEM_{00} mode are known, all higher order modes simply scale from it in a known manner. Diffraction effects due to the finite size of the mirrors will be neglected in this section.

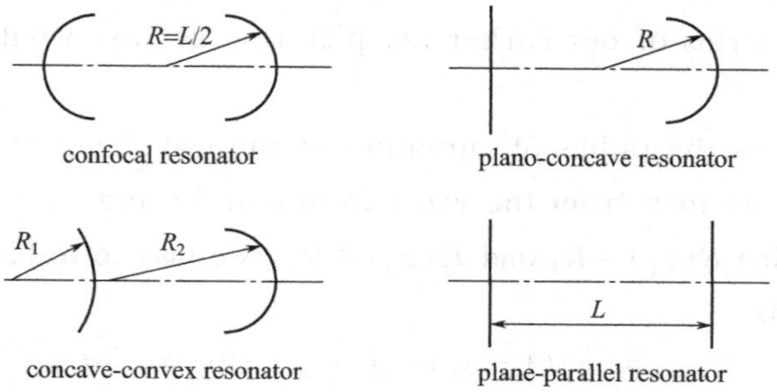

Figure 5-5 Common resonator configurations

Eq. (5. 17) ~ Eq. (5. 19) treat the most general case of a resonator. There are many optical resonator configurations for which Eq. (5. 17) ~ Eq. (5. 19) are greatly simplified.

(1) Mirrors of equal curvature

A special case of a symmetrical configuration is the concentric resonator that consists of two mirrors separated by twice their radius, that is, $R = L/2$. The corresponding beam consists of a mode whose dimension is fairly large at each mirror and focuses down to a diffraction-limited point at the center of the resonator. A concentric resonator is rather sensitive to misalignment, and the small spot can lead to optical damage.

Another very important special case of a resonator with mirrors of equal curvature is the confocal resonator. For this type of resonator the mirror separation equals the curvature of the identical mirrors, that is, $R = L$. The confocal configuration gives the smallest possible mode dimension for a resonator of given length. For this reason, confocal resonators are not often employed since they do not make efficient use of the active material.

(2) Plano-concave resonator

For a resonator with one flat mirror ($R_1 = \infty$) and one curved mirror, it can be found that the beam waist w_0 occurs at the flat mirror.

A special case of this resonator configuration is the hemispherical resonator. The hemispherical resonator consists of one spherical mirror and one flat mirror placed approximately at the center of curvature of the sphere. The resultant mode has a relatively large diameter at the spherical mirror and focuses to a diffraction-limited point at the plane mirror. In practice, one makes the mirror separation d slightly less than R_2 so that a value of w_1 is obtained that gives reasonably small diffraction losses.

In solid-state lasers, the small spot size can lead to optical damage at the mirror. A near hemispherical resonator has the best alignment stability of any configuration; therefore it is often employed in low-power lasers such as He-Ne lasers.

(3) Concave-convex resonator

A small radius convex mirror in conjunction with a large-radius concave or plane mirror is a very common resonator in high-average-power solid-state lasers. As follows from the discussion in the next section, as a passive resonator such a configuration is unstable.

However, in a resonator that contains a laser crystal, this configuration can be stable since the diverging properties of the convex mirror are counteracted by the focusing action of the laser rod. Since the convex mirror partially compensates for thermal lensing, a large mode volume can be achieved.

(4) Plane-parallel resonator

The plane-parallel or flat-flat resonator, which can be considered a special case of the large-radius mirror configuration if $R_1 = R_2 = \infty$, is extremely sensitive to perturbation. However, in an active resonator, that is, a resonator containing a laser crystal, this configuration can be quite useful. Heat extraction leads to thermal lensing in the active medium; this internal lens has the effect of transforming the plane-parallel resonator to a curved mirror configuration. Therefore, the thermally induced lens in the laser material brings the flat-flat resonator into geometric stability.

5. 2. 5 Stability of Laser Resonators

The expressions for $w(z_1)$ and $w(z_2)$ contain a negative root of $(1-g_1g_2)$. Unless the product g_1g_2 is less than 1, the spot sizes on the mirrors become infinite or imaginary. Laser cavities for which g_1g_2 exceeds 1 are unstable; those for which the product is just less than 1 are on the border of stability, because the spot size may exceed the mirror size and bring about great loss. Thus the stability criterion for lasers is

$$g_1g_2=\left(1-\frac{d}{R_1}\right)\left(1-\frac{d}{R_2}\right)<1 \tag{5.21}$$

The limiting case, $g_1g_2=1$, is a hyperbola. Stable resonators lie between the two branches of the hyperbola, unstable resonators, outside the two branches. Figure 5-6 shows the stability diagram for laser resonator. Resonators that are least sensitive to changes within the cavity (such as thermally induced focusing effects in solid or liquid lasers) lie on the hyperbola $g_1g_2=1/2$.

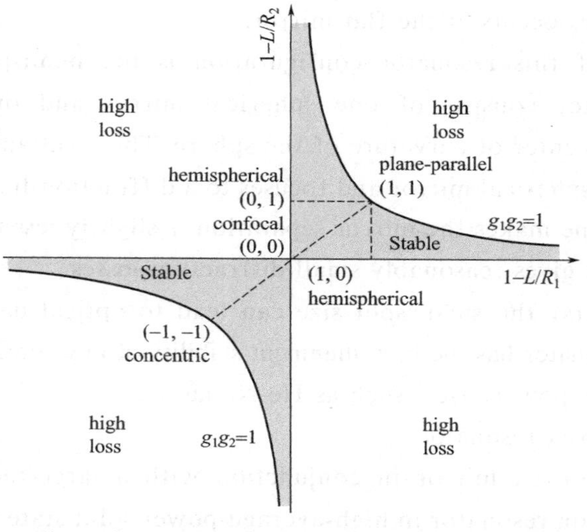

Figure 5-6 Stability diagram for laser resonator

5. 3 Laser Amplifiers

The power or energy from an oscillator with specific spatial, temporal, or spectral properties can be increased by adding one or more amplifying stages to the laser system. The main function of the amplifier is to increase the brightness of the beam. Amplification of pulsed or continuous wave (CW) power from an oscillator can be achieved using one of several techniques, including a master-oscillator power amplifier (MOPA) concept, a regenerative amplifier, a seeded or injection-locked power oscillator, and series connection of several gain modules within a common resonator. The choice of the specific amplifier approach depends on the power level, spectral and temporal characteristics of the input signal, as well as the desired output energy or power.

5.3.1 Pulse Amplification

The use of lasers as pulse amplifiers is of great interest in the design of high-energy, high brightness radiation sources. In the pulse amplifiers, the input Q-switched or mode-locked pulse is considerably shorter than the fluorescent lifetime of the active medium. Hence the effect of spontaneous emission and pumping rate on the population inversion during the amplification process can be neglected. Furthermore, energy is extracted from the amplifier, which was stored in the amplifying medium, prior to the arrival of the pulse.

Consider the one-dimensional case of a beam of monochromatic radiation incident on the front surface of an amplifier rod of length l. The point at which the beam enters the gain medium is designated the reference point, $x=0$.

If we ignore the effect of fluorescence and pumping during the pulse duration, we obtain for the population inversion

$$\partial n/\partial t = -\gamma n c \sigma \Phi \tag{5.22}$$

where $\gamma = 1 + g_2/g_1$ (g_1 and g_2 are the degeneracy of energy levels) for a three level system and $\gamma = 1$ for a four level system, c is the speed of light in the medium, σ denotes the emission cross section, and Φ represents the photon density.

The growth of a pulse traversing a medium with an inverted population is described by the nonlinear, time-dependent photon-transport equation, which accounts for the effect of the radiation on the active medium and vice versa.

$$\frac{\partial \Phi}{\partial t} = c n \sigma \Phi - \frac{\partial \Phi}{\partial x} c \tag{5.23}$$

The rate at which the photon density changes in a small volume of material is equal to the net difference between the generation of photons by the stimulated emission process and the flux of photons which flows out from that region.

The two differential equations [Eq. (5.22) and Eq. (5.23)] must be solved for the inverted electron population n and the photon flux Φ. Frantz and Nodvik solved these nonlinear equations for various types of input pulse shapes. The results are

$$E_{\text{out}} = E_s \ln\left\{1 + \left[\exp\left(\frac{E_{\text{in}}}{E_s}\right) - 1\right] \exp(g_0 l)\right\} \tag{5.24}$$

where E_{in} and E_{out} are input and output fluence (energy per unit area), respectively, and the saturation fluence E_s can be defined by

$$E_s = \frac{h\nu}{\gamma\sigma} = \frac{E_{\text{st}}}{\gamma g_0} \tag{5.25}$$

where E_{st} is the stored energy per volume, and g_0 is the small-signal gain coefficient.

The gain of the amplifier can be expressed as

$$G = \frac{E_s}{E_{in}} \ln\left\{1 + \left[\exp\left(\frac{E_{in}}{E_s}\right) - 1\right] G_0\right\} \tag{5.26}$$

Eq. (5.26) permits one to calculate the gain of an amplifier as a function of the input energy density, provided that the small-signal gain or the energy stored in the amplifier is known.

In a laser system which has multiple stages, these equations can be applied successively, where by the output of one stage becomes the input for the next stage.

5.3.2 Signal Distortions

As an optical signal propagates through a laser amplifier, distortions will arise as the result of a number of physical processes. We can distinguish between spatial and temporal distortions.

(1) Spatial distortions

The development of high-power laser oscillator-amplifier systems has led to considerable interest in the quality of the output beam attainable from these devices. Beam distortions produced during amplification may lead either to an increase in divergence or to localized high energy densities, which can cause laser rod damage. We will consider the main phenomena producing a spatial distortion, which is analogous to a wavefront or intensity distortion. The main factors producing the spatial distortions are as follows: ①Nonuniform pumping. Because of the exponential absorption of pump light, the center of the rod is pumped less than the edges, resulting in the spatial distortion of the signal. ② Non-uniformities in the active material. Even laser rods with excellent optical quality contain a small amount of inherent stress, index of refraction variations, gradients in the active ion concentration, contaminants, inclusions, etc.. These non-uniformities will significantly modify the energy distribution of an incoming signal. ③ Gain saturation. A beam propagating through an amplifier can experience a distortion of the spatial profile because of the saturation-induced change in the distribution of gain. The weaker portions of the signal are amplified relatively more than the stronger portions because they saturate the medium to a lesser degree. ④ Diffraction effects. Any limiting aperture in the amplifier section that removes energy at the edges of the beam will give rise to Fresnel rings. Diffraction effects are undesirable because they introduce strong intensity modulation in the beam that can produce spatial hot spots and lead to optical damage. ⑤ Thermal distortions. The nonuniform pumping which leads to a higher gain coefficient at the edges of a laser rod also causes a nonuniform temperature profile across the rod. The absorbed pump power raises the temperature of the surface of the rod above the temperature of the rod center. As a result, a negative thermal lens is created, which distorts the wavefront of the beam.

(2) Temporal distortions

For a square pulse traversing an amplifier, the leading pulse edge sees a larger inverted population than the trailing edge. This occurs simply because the leading edge stimulates the release of some of the stored energy and decreases the population inversion for the trailing edge. Thus, less energy is added to the final portions of a pulse than to the leading regions. The pulse-shape evolution depends appreciably on the rise time and shape of the leading edge of the input pulse. In general, one observes a forward shift of the peak as the pulse propagates through the amplifying medium.

Figure 5-7 illustrates the change of the pulse shape in an amplifier as a result of gain saturation. Curve 1 is the initially rectangular pulse, curves 2 and 3 are for $g_0 = 0.1\ \text{cm}^{-1}$, $l = 20$ cm, $E_{in}/E_s = 0.5$, and $E_{in}/E_s = 0.1$, respectively.

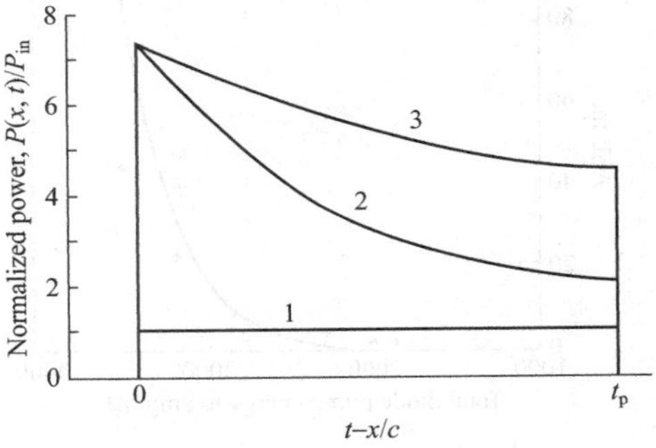

Figure 5-7 Change of the pulse shape in an amplifier as a result of gain saturation

The intensity-induced index changes, are not only responsible for the catastrophic self-focusing problems in high-power laser amplifiers, but also cause a frequency shift of the pulse in the amplifying medium. In mode-locked pulses, the frequency modulation causes a broadening of the pulse width.

5. 3. 3 Amplified Spontaneous Emission

The level of population inversion which can be achieved in an amplifier is usually limited by depopulation losses which are caused by amplified spontaneous emission (ASE). The favorable condition for strong ASE is a high gain combined with a long path length in the active material. In these situations, spontaneous decay occurring near one end of a laser rod can be amplified to a significant level before leaving the active material at the other end. A threshold for ASE does not exist; however, the power emitted as fluorescence increases rapidly with gain. Therefore, as the pump power increases, ASE becomes the dominant decay mechanism for the laser amplifier.

In high-gain, multistage amplifier systems, ASE may become large enough to deplete the upper state inversion. ASE is particularly important in large Nd: glass systems employed in fusion experiments. An analytical expression for the fluorescence flux I_{ASE} from a laser rod as a function of small signal gain, which has been found very useful in estimating the severity of ASE, is given in Eq. (5. 27) with approximation $G_0 > 1$.

$$\frac{I_{ASE}}{I_s} = \frac{\Omega}{4} \frac{G_0}{(\ln G_0)^{1/2}} \tag{5.27}$$

where I_s is the saturation flux, Ω is the solid angle around the axis of the active material and G_0 is the small signal gain of the active medium.

In Figure 5-8, the measured ASE from a four-stage, double-pass Nd: YAG amplifier chain is plotted as a function of diode-pump input, in which dots are measured values and solid line represents calculated values from Eq. (5. 27). It can be seen that at an input of about 500 mJ into each amplifier, ASE starts to become noticeable, and quickly increases in intensity to reach 75 mJ at an input of 900 mJ per amplifier.

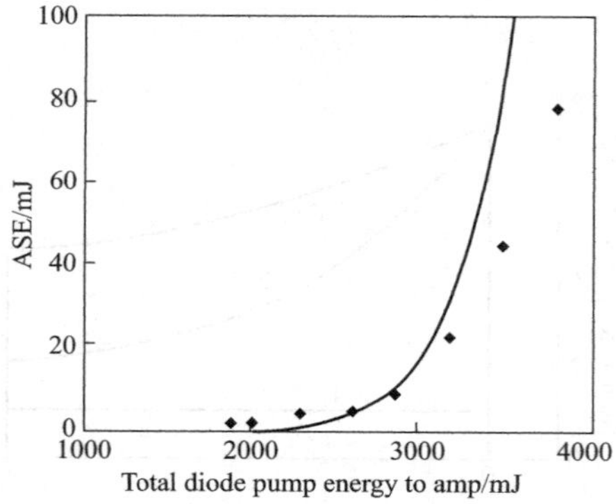

Figure 5-8 ASE from a 4-stage double pass Nd: YAG amplifier chain

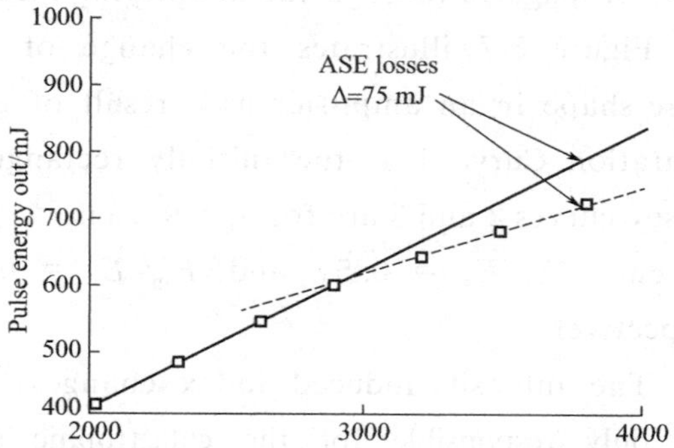

Figure 5-9 Signal output versus pump input from a multistage Nd: YAG amplifier chain

The detrimental effect of ASE can be seen in Figure 5-9, which shows the output from the amplifier chain as a function of pump input. As the amplified spontaneous emission increases, the slope of output versus input decreases and the difference can be accounted for by the loss due to ASE.

So far, we discussed the amplification of fluorescence generated by the active material itself. It has been observed that flashlamp pump radiation, which is within the spectral region of the laser transition, will also be amplified and can lead to a reduction of stored energy in the laser material.

Gain saturation resulting from amplification of pump radiation in the active medium has been observed mainly in Nd: glass and Nd: YAG amplifiers. In addition to the depumping process caused by the lamp radiation, fluorescence being emitted from the laser rod escaping out into the laser pump cavity can be reflected back into the laser rod and thereby stimulating further off-axis emission. Thus the presence of pump-cavity walls, which have a high effective reflectivity at the fluorescence wavelength, causes a transverse depumping action which depletes the energy available for on-axis stimulated emission. Elimination of these effects may require the use of optical filters in the pump cavity, cladding of the laser rod with a material which absorbs at the laser wavelength, or the addition of chemicals to the cooling fluid which serve the same purpose.

5.4 Laser Techniques

5.4.1 Q-Switching

A mode of laser operation extensively employed for the generation of high pulse power is known as Q-switching. Laser resonators are characterized by the quality factor Q, which is defined as the ratio of the energy stored in the cavity to the energy loss per cycle, i. e.,

$$Q = 2\pi\left[1 - \exp\left(-\frac{2\pi\omega_0}{\tau_c}\right)\right]^{-1} \tag{5.28}$$

where ω_0 denotes the angular frequency , τ_c represents all the losses in an optical resonator of a laser oscillator. Since τ_c has the dimension of time, the losses are expressed in terms of a relaxation time.

In the technique of Q-switching, energy is stored in the amplifying medium by optical pumping while the cavity Q is lowered to prevent the onset of laser emission. Although the energy stored and the gain in the active medium are high, the cavity losses are also high, lasing action is prohibited, and the population inversion reaches a level far above the threshold for normal lasing action. The time for which the energy may be stored is on the order of the lifetime of the upper level of the laser transition. When a high cavity Q is restored, the stored energy is suddenly released in the form of a very short pulse of light. Because of the high gain created by the stored energy in the active material, the excess excitation is discharged in an extremely short time. The peak power of the resulting pulse exceeds that obtainable from an ordinary long pulse by several orders of magnitude. The development of a Q-switched laser pulse is shown in Figure 5-10, in which the flashlamp output, resonator loss, population inversion, and photon flux is a function of time.

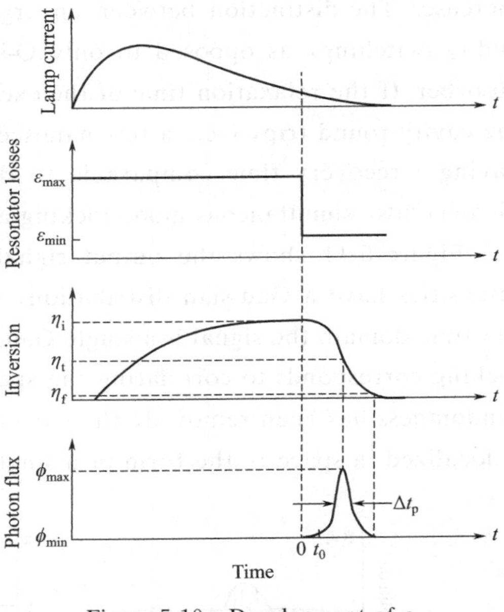

Figure 5-10 Development of a Q-switched laser pulse

Q-switches have been designed based upon rotational, oscillatory, or translational motion of optical components. These techniques have in common that they inhibit laser action during the pump cycle by either blocking the light path, or causing a mirror misalignment, or reducing the reflectivity of one of the resonator mirrors. Near the end of the pumping pulse, when maximum energy has been stored in the laser medium, a high Q-condition is established and a Q-switch pulse is emitted from the laser.

5.4.2 Mode Locking

The nonlinear absorption of saturable absorbers was first successfully employed for simultaneously Q-switching and mode-locking solid-state lasers in 1965. The saturable absorbers consisted of organic dyes that absorb at the laser wavelength. At sufficient intense laser radiation, the ground state of the dye becomes depleted, which decreases the losses in the resonator for increasing pulse intensity.

In pulsed mode-locked solid-state lasers, pulse shortening down to the limit set by the gain-bandwidth is prevented because of the early saturation of the absorber which is a result of the simultaneously occurring Q-switching process. Shorter pulses and a much more reproducible performance are obtained if the transient behavior due to Q-switching is eliminated. In steady-state or CW mode locking, components or effects are utilized which exhibit a saturable absorber-like behavior, i. e., a loss that decreases as the laser intensity increases. The distinction between an organic dye suitable for simultaneous mode-locking and Q-switching, as opposed to only Q-switching the laser, is the recovery time of the absorber. If the relaxation time of the excited-state population of the dye is on the order of the cavity round trip, i. e., a few nanoseconds, passive Q-switching will occur. With a dye having a recovery time comparable to the duration of mode-locked pulses, i. e., a few picoseconds, simultaneous mode-locking and Q-switching can be achieved.

Figure 5-11 shows the output signal of an ideally mode-locked laser. The spectral intensities have a Gaussian distribution, while the spectral phases are identically zero. In the time domain the signal is a single Gaussian pulse. As can be seen from this figure, mode locking corresponds to correlating the spectral amplitudes and phases. When all the initial randomness has been removed, the correlation of the modes is complete and the radiation is localized in space in the form of a single pulse.

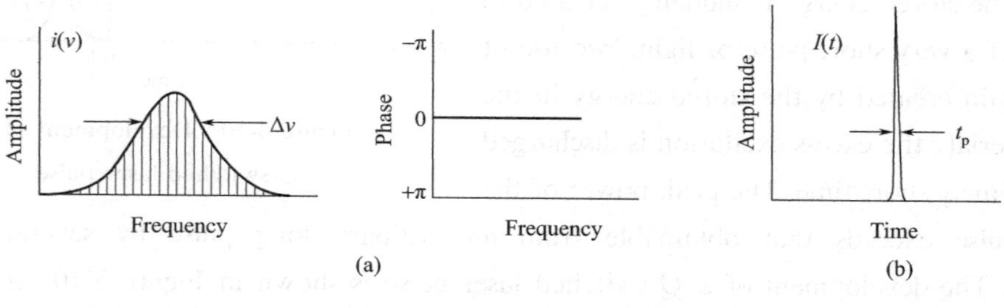

(a) the spectral intensities have a Gaussian distribution, while the spectral phases are identically zero;

(b) in the time domain the signal is a transform-limited Gaussian pulse.

Figure 5-11 Signal structure of an ideally mode-locked laser

In recent years, several passive mode-locking techniques have been developed for solid-state lasers whereby fast saturable absorber-like action is achieved in solids. Most of these novel optical modulators utilize the nonresonant Kerr effect. The Kerr effect produces intensity-dependent changes of the refractive index. It is generally an undesirable effect because it can lead to self focusing and filament formation in intense beams. In contrast to the absorption in bleachable dyes, the nonresonant Kerr effect is extremely fast, wavelength-independent, and allows the generation of a continuous train of mode-locked pulse from a CW-pumped laser.

5.4.3 Mode Selecting

(1) Transverse modes

Many applications of lasers require the operation of laser at the TEM_{00} mode, since this mode produces the smallest beam divergence, the highest power density, and, hence, the highest brightness. Focusing a fundamental-mode beam by an optical system will produce a diffraction-limited shot of maximum power per unit area. Generally speaking, in many applications it is a high brightness rather than large total emitted power that is desired from the laser.

Furthermore, the radiation profile of the TEM_{00} mode is smooth. This property is particularly important at higher power levels, since multimode operation leads to the random occurrence of local maxima in intensity, so-called hot spots, which can exceed the damage threshold of the optical components in the resonator.

Transverse mode selection generally restricts the area of the laser cross section over which oscillation occurs, thus decreasing the total output power. However, mode selection reduces the beam divergence so that the overall effect of mode selection is an increase in the brightness of the laser.

Most lasers tend to oscillate not only in higher-order transverse modes, but in many such modes at once. Because of the fact that higher-order transverse modes have a larger spatial extent than the fundamental mode, a given size aperture will preferentially discriminate against higher-order modes in a laser resonator. As a result, the question of whether or not a laser will operate in the lowest-order mode depends on the size of this mode and the diameter of the smallest aperture in the resonator. If the aperture is much smaller than the TEM_{00} mode size, large diffraction losses will occur which will prevent the laser from oscillating. If the aperture is much larger than the TEM_{00} mode size, then the higher-order modes will have sufficiently small diffraction losses to be able to oscillate.

(2) Longitudinal modes

A typical laser will oscillate in a band of discrete frequencies which have an overall width of about 10^{-4} of the laser frequency. Although this is a rather monochromatic light source, there are still many applications for which greater spectral purity is required.

In one of the earliest attempts to narrow the spectral width of a laser, tilted Fabry-Perot etalons were employed as mode-selecting elements. Also the concept of axial mode selection based on an analysis of the modes of a multiple-surface resonator was introduced. Since then, many mode-selecting techniques have been developed, such as interferometric mode selection, enhancement of longitudinal mode selection, injection seeding etc..

① Interferometric mode selection. A Fabry-Perot-type reflector is inserted between the two mirrors of the optical resonator. This will cause a strong amplitude modulation of the closely spaced reflectivity peaks of the basic laser resonator and thereby prevent most modes from reaching threshold.

② Enhancement of longitudinal mode selection. In this case, an inherent mode-selection process in the resonator is further enhanced by changing certain system parameters, such as shortening the resonator, removal of spatial hole burning or lengthening of the Q-switch built-up time.

③ Injection seeding. This technique takes advantage of the fact that stable, single longitudinal mode operation can readily be achieved in a very small crystal located within a short, traveling wave resonator. These devices which are end-pumped by a laser diode array are by themselves not powerful enough for most applications, unless the output of such a device is coupled into a large slave oscillator for amplification.

5.4.4 Frequency Control

Frequency control requires a certain amount of tunability of the laser emissions, either to lock the frequency to a reference or to an incoming signal for coherent detection, or to provide an output signal at a specific wavelength for spectroscopic studies.

In certain applications, such as coherent light detection and ranging (LIDAR) system, stabilized single-frequency laser source is required, both as local oscillator for the detection system, and as injection seeding source for a pulsed transmitter.

Miniature diode-pumped solid-state lasers, such as monolithic or microchip devices, are particularly attractive for this application because of their single-frequency operation. Furthermore, the short resonator of these devices supports a relatively large tuning range because longitudinal modes are widely separated. In monolithic lasers, the frequency can be tuned by thermal expansion or by mechanical stress applied to the crystal. Discrete-element miniature lasers can be tuned by insertion of an electro-optic modulator into the cavity or by cavity-length adjustment.

In monolithic lasers such as the microchip laser, or nonplanar ring laser, the resonator mirrors are coated directly on the laser crystal, which precludes the introduction of traditional intra-cavity elements. However, varying the temperature of the crystal will change the resonator length, and therefore frequency tunes the laser emission.

Instead of changing the temperature of the laser crystal by means of a TE cooler, pump power modulation has been successfully employed to cause rapid frequency tuning in miniature lasers. As the pump power increases, more thermal energy is deposited in the gain medium, raising its temperature and therefore changing the resonant frequency of the laser cavity. Modulation of the pump power has the undesirable effect of changing the laser output power. In implementing this technique, TE coolers maintain a fixed average temperature of the crystal. Around this temperature point, rapid thermal changes are induced by means of current modulation of the pump-laser diode. The thermal response is sufficiently fast, permitting phase locking of the laser.

Another technique for precisely tuning the laser frequency of a monolithic laser is based on stress-induced changes in the resonator. In these designs, stress is applied to the laser crystal by using a piezoelectric transducer. This results in a stress-induced birefringence of the refractive index, and a strain-related distortion or elongation of the resonator.

In the case of discrete-element microcavity lasers, frequency tuning has been achieved by cavity length modulation or by incorporating an electro-optic modulator into the resonator. In the former design, an air gap between the two pieces forming a ring resonator was adjusted by a piezoelectric transducer. Cavity length tuning was achieved over a frequency range of 13.5 GHz at a speed on the order of milliseconds.

Frequency stability is degraded by mechanical vibrations, acoustic noise, and ambient temperature variations. The most stringent requirement for absolute frequency controls is imposed on scientific lasers developed for gravitational-wave detectors. Absolute frequency stability can be achieved by locking the output of the laser to a high finesse reference Fabry-Perot cavity employing the Pound-Drever-Hall (PDH) servo system.

5.4.5 Wavelength Selection

In the operation of lasers with very broad gain curves, such as alexandrite, Ti: sapphire, and Cr: GSGG, it is necessary to use a wavelength selection technique to ①restrict laser action to a specified wavelength and ② tune the laser output. Several different methods are available (in principle) for providing the wavelength selection and tuning. These include ① use of a prism inside the resonator, ② utilization of an adjustable optical grating within the laser, ③ use of intracavity etalons, or ④ use of one or more thin birefringent plates within the laser that are tilted at Brewster's angle.

The technique most commonly employed for the wavelength selection of tunable lasers is the birefringent filter. In its simplest form, the birefringent filter consists of a single thin birefringent crystal located inside the laser, as shown in Figure 5-12.

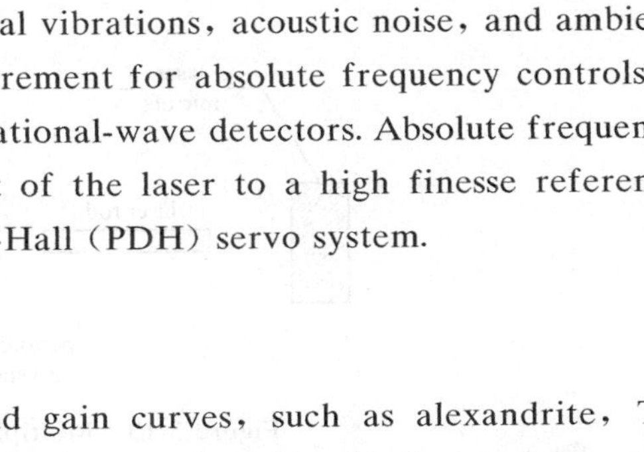

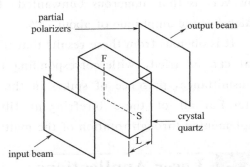

Figure 5-12 Single-crystal wavelength selector

For simplicity, we assume that the birefringent axes lie in the plane of the crystal, and that the crystal is tilted at Brewster's angle. Wavelength selection occurs with the birefringent filter because of the two different crystal indices of refraction. When the laser light has a wavelength corresponding to an integral number of full-wave retardations, the laser operates as if the filter were not present. At any other wavelength, however, the laser mode polarization is modified by the filter and suffers losses at the Brewster surfaces.

Tunability of the laser is achieved by rotating the birefringent crystal in its own plane. This changes the included angle φ between crystal optic axis and the laser axes and, hence, the effective principal refractive indices of the crystal. The amplitude transmittance of the single-stage filter of Figure 5-12 has been calculated. For a quartz crystal rotated to $\varphi = 45^\circ$ and tilted to Brewster's angle, the transmittance at unwanted wavelengths is about 82%. This may or may not provide adequate suppression of unwanted wavelengths for certain lasers. One way to lower the filter transmittance in the rejection band is to use a stack of identical crystal plates that are similarly aligned, as shown in Figure 5-13. If one uses a stack of 10 quartz plates, the resulting transmittance in the rejection band is about 15%, which is certainly small enough to suppress unwanted laser frequencies.

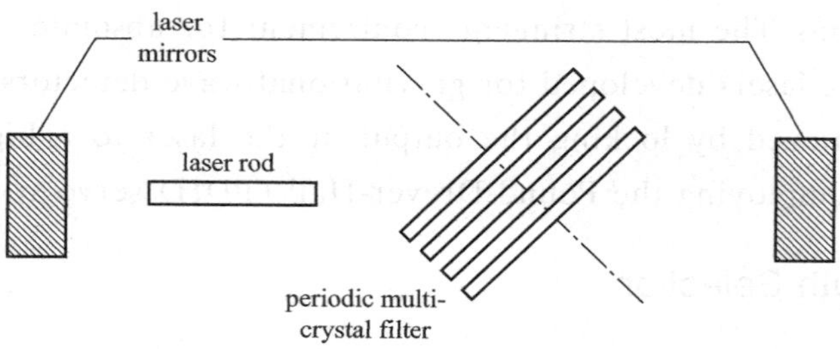

Figure 5-13　Multiple-crystal wavelength selector

Another technique for lowering the filter transmittance in the rejection band is to include more Brewster's angle surfaces in the laser. Still another approach for narrowing the width of the central passband of the filter is to use several crystals in series whose thicknesses vary by integer ratios. The central passband of this kind of filter is considerably narrower than those of the previously discussed designs. The disadvantage of this approach, however, is that numerous (unwanted) transmission spikes are present, the largest of which has the amplitude of about 75%.

It is obvious from these results that there are virtually an unlimited number of designs that can be tried, with corresponding tradeoffs in central passband width, stopband transmittance, presence of spikes in the stopband, and complexity of the birefringent filter. For all of these birefringent filter designs, tuning is continuous and easily implemented through rotation of the multiple crystals.

5.5 Laser Applications

When lasers were invented in 1960, they were called "a solution looking for a problem". Since then, they have become ubiquitous, finding utility in thousands of highly varied applications in every field of modern society, including consumer electronics, information technology, science, medicine, industry, law enforcement, entertainment, and military, etc.. In this section, we will briefly introduce laser applications in four fields: military, medical treatment, industry, and laser cooling.

5.5.1 Laser in Military

Military uses of lasers include applications such as lidar and rangefinder, target designation and ranging, defensive countermeasures, communications and directed energy weapons. Directed energy weapons are also in use, such as Boeing's Airborne Laser which is constructed inside a Boeing 747. It disrupts the trajectory of shoulder-fired missiles.

(1) Lidar

Light detection and ranging (LIDAR, also LADAR) is an optical remote sensing technology that can measure the distance to, or other properties of a target by illuminating the target with light, often using pulses from a laser. LIDAR technology has application in archaeology, geography, geology, geomorphology, seismology, forestry, remote sensing and atmospheric physics, as well as in airborne laser swath mapping (ALSM), laser altimetry and LIDAR contour mapping.

LIDAR has been used extensively for atmospheric research and meteorology. Downward looking LIDAR instruments fitted to aircraft and satellites are used for surveying and mapping, a recent example is the National Aeronautics and Space Administration (NASA) experimental advanced research lidar. In addition, LIDAR has been identified by NASA as a key technology for enabling autonomous precision safe landing of future robotic and crewed lunar landing vehicles.

LIDAR uses ultraviolet, visible, or near infrared light to image objects and can detect a wide range of targets, including non-metallic objects, rocks, rain, chemical compounds, aerosols, clouds and even single molecules. A narrow laser beam can be used to map physical features with very high resolution.

In general, there are two kinds of lidar detection schema: incoherent or direct energy detection (which is principally an amplitude measurement) and coherent detection (which is best for doppler, or phase sensitive measurements). Coherent systems generally use optical heterodyne detection which is more sensitive than direct detection, that allows them to operate a much lower power but at the expense of more complex transceiver requirements.

In both coherent and incoherent lidar, there are two types of pulse models: micro pulse lidar systems and high energy systems. Micro pulse systems have developed as a result of the ever increasing amount of computer power available combined with advances in laser technology. They use considerably less energy in the laser, typically on the order of one micro joule, and are often "eye-safe", meaning they can be used without safety precautions. High-power systems are common in atmospheric research, where they are widely used for measuring many atmospheric parameters: the height, layering and densities of clouds, cloud particle properties (extinction coefficient, backscatter coefficient, depolarization), temperature, pressure, wind, humidity, trace gas concentration (ozone, methane, nitrous oxide, etc.).

There are several major components to a lidar system:

① Laser. 600～1000 nm lasers are most common for non-scientific applications. They are inexpensive, but since they can be focused and easily absorbed by the eye, the maximum power is limited by the need to make them eye-safe. Eye-safety is often a requirement for most applications.

② Scanner and optics. How fast images can be developed is also affected by the speed at which they are scanned. There are several options to scan the azimuth and elevation, including dual oscillating plane mirrors, a combination with a polygon mirror, a dual axis scanner. Optic choices affect the angular resolution and range that can be detected. A hole mirror or a beam splitter are options to collect a return signal.

③ Photodetector and receiver electronics. Two main photodetector technologies are used in lidar: solid state photodetectors, such as silicon avalanche photodiodes, or photomultipliers. The sensitivity of the receiver is another parameter that has to be balanced in a LIDAR design.

④ Position and navigation systems. LIDAR sensors that are mounted on mobile platforms such as airplanes or satellites require instrumentation to determine the absolute position and orientation of the sensor. Such devices generally include a global positioning system (GPS) receiver and an inertial measurement unit (IMU).

(2) Laser rangefinder

A laser rangefinder is a device which uses a laser beam to determine the distance to an object. The most common form of laser rangefinder operates on the time of flight principle by sending a laser pulse in a narrow beam towards the object and measuring the time taken by the pulse to be reflected off the target and returned to the sender. Due to the high speed of light, this technique is not appropriate for high precision sub-millimeter measurements, where triangulation and other techniques are often used.

Rangefinders provide an exact distance to targets located beyond the distance of point-blank shooting to snipers and artillery. They also can be used for military reconciliation and engineering.

Handheld military rangefinders operate at ranges of 2 km up to 25 km and are combined with binoculars or monoculars. When the rangefinder is equipped with a digital magnetic compass (DMC) and inclinometer, it is capable of providing magnetic azimuth, inclination, and height (length) of targets. Some rangefinders can also measure the speed of the target in relation to the observer. Some rangefinders have cable or wireless interfaces to enable them to transfer their measurement data to other equipment like fire control computers. Some models also offer the possibility to use add-on night vision modules. Most handheld rangefinders use standard or rechargeable batteries.

The more powerful models of rangefinders measure distance up to 25 km and are normally installed either on a tripod or directly on a vehicle or gun platform. In the latter case the rangefinder module is integrated with on-board thermal, night vision and daytime observation equipment. The most advanced military rangefinders can be integrated with computers.

To make laser rangefinders and laser-guided weapons less useful against military targets, various military arms may have developed laser-absorbing paint for their vehicles. Regardless, some objects do not reflect laser light very well and using a laser rangefinder on them is difficult.

(3) Defensive countermeasure

Defensive countermeasure applications can range from compact, low power infrared countermeasures to high power, airborne laser systems. IR countermeasure systems use lasers to confuse the seeker heads on heat-seeking anti-aircraft missiles. High power boost-phase intercept laser systems use a complex system of lasers to find, track and destroy intercontinental ballistic missiles (ICBM). In this type of system a chemical laser, in which the laser operation is powered by an energetic chemical reaction, is used as the main weapon beam. The mobile tactical high-energy laser (MTHEL) is another defensive laser system under development; this is envisioned as a field-deployable weapon system able to track incoming artillery projectiles and cruise missiles by radar and destroy them with a powerful deuterium fluoride laser.

Another example of direct use of a laser as a defensive weapon was researched for the strategic defense initiative (SDI, nicknamed "Star Wars"), and its successor programs. This project would use ground-based or space-based laser systems to destroy incoming ICBMs. The practical problems of using and aiming these systems were many; particularly the problem of destroying ICBMs at the most opportune moment, the boost phase just after launch. This would involve directing a laser through a large distance in the atmosphere, which, due to optical scattering and refraction, would bend and distort the laser beam, complicating the aiming of the laser and reducing its efficiency.

Another idea from the SDI project was the nuclear-pumped X-ray laser. This was essentially an orbiting atomic bomb, surrounded by laser media in the form of glass rods; when the bomb exploded, the rods would be bombarded with highly-energetic gamma-ray photons, causing spontaneous and stimulated emission of X-ray photons in the atoms making up the rods. This would lead to optical amplification of the X-ray photons, producing an X-ray laser beam that would be minimally affected by atmospheric distortion and capable of destroying ICBMs in flight. The X-ray laser would be a strictly one-shot device, destroying itself on activation. Some initial tests of this concept were performed with underground nuclear testing; however, the results were not encouraging. Research into this approach to missile defense was discontinued after the SDI program was cancelled.

(4) Target designator

Another military use of lasers is as a laser target designator. This is a low-power laser pointer used to indicate a target for a precision-guided munition, typically launched from an aircraft. The guided munition adjusts its flight-path to home in to the laser light reflected by the target, enabling a great precision in aiming. The beam of the laser target designator is set to a pulse rate that matches that set on the guided munition to ensure munitions strike their designated targets and do not follow other laser beams which may be in use in the area. The laser designator can be shined onto the target by an aircraft or nearby infantry. Lasers used for this purpose are usually infrared lasers, so the enemy cannot easily detect the guiding laser light.

(5) Laser sight

Laser has been used in most firearms applications as a tool to enhance the targeting of other weapon systems. For example, a laser sight is a small, usually visible-light laser placed on a handgun or a rifle and aligned to emit a beam parallel to the barrel. Since a laser beam has low divergence, the laser light appears as a small spot even at long distances; the user places the spot on the desired target and the barrel of the gun is aligned (but not necessarily allowing for bullet drop, windage and the target moving while the bullet travels).

Most laser sights use a red laser diode. Others use an infrared diode to produce a dot invisible to the naked human eye but detectable with night vision devices. In the late 1990s, green diode pumped solid state laser (DPSS) sights (532 nm) became available. Modern laser sights are small and light enough for attachment to the firearms.

In 2007, Laser Max, a company specializing in manufacturing lasers for military and police firearms, introduced the first mass-production green laser available for small arms. This laser mounts to the underside of a handgun or long arm on the accessory rail. The green laser is supposed to be more visible than the red laser in bright lighting conditions because, for the same wattage, green light appears brighter than red light.

(6) Eye-targeted lasers

A non-lethal laser weapon was developed by the U. S. Air Force to temporarily impair an adversary's ability to fire a weapon or to otherwise threaten enemy forces. This unit illuminates an opponent with harmless low-power laser light and can have the effect of dazzling or disorienting the subject or causing him to flee. Several types of dazzlers are now available, and some have been used in combat.

There remains the possibility of using lasers to blind, since this requires much lower power levels, and is easily achievable in a man-portable unit. However, most nations regard the deliberate permanent blinding of the enemy as forbidden by the rules of war. Although several nations have developed blinding laser weapons, such as China's ZM-87, none of these are believed to have made it past the prototype stage.

In addition to the applications that crossover with military applications, a widely known law enforcement use of lasers is for lidar to measure the speed of vehicles.

5. 5. 2 Laser in Medicine

Medicine and the biosciences are important fields for implementing lasers. Even before early lasers had left research laboratories for industrial applications, medicine was considered an important "consumer" of laser technologies, first as manufacturing tools for medical instruments, then as working instruments themselves. Practically all types of lasers have found their specific niches in important branches of medicine: research, monitoring, imaging, probing, therapy, surgery, and others. Referring to more specific applications, lasers are used literally everywhere. In biomedical investigation: fluorescent spectroscopy, microscopy, and flow cytometry. In surgery: "bloodless" operations in cardiology, on abdominal and thoracic organs, and skull and brain microsurgery. In cosmetic and aesthetic medicine: smoothing wrinkles, resurfacing the skin, and bleaching tattoos. In therapy: the treatment of cancer, spider veins, and vascular dysfunction. In diagnostics: endoscopic investigations and optical coherence tomography (OCT). This list can be extended further by going deeper into subclassifications and interdisciplinary topics.

Specially, fiber lasers should be considered successors of trends in medical applications rather than "pioneers" discovering untouched fields. However, due to their inherent flexibility of physical principles and design, as well as outstanding performance, fiber lasers have enormous potential to bring new opportunities to medicine.

Before specifying particular applications of fiber lasers in medicine, it is instructive to briefly depict some of the main fields in which lasers are commonly used for health care, monitoring, and research regardless of which particular type of laser is considered. Such a classification is commonly arranged according to how organic tissues react to laser radiation.

Using the terminology adopted among medical professionals, the reaction of organic tissues to laser radiation is typically described as an optical or thermal response, as shown in Figure 5-14.

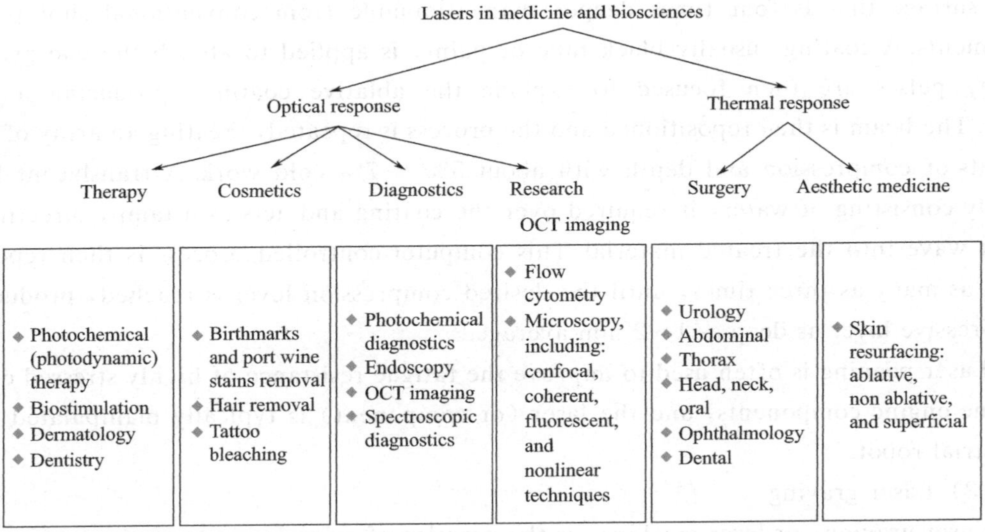

Figure 5-14 General applications of laser in medicine and life sciences

In the optical response, the light energy absorbed does not damage or destroy the tissues. Most effects are achieved either ① by selective resonance absorption of specific laser wavelengths by fluorophores or photosensitizers with sequentially photoinduced changes of the tissues, or ② by exposure with short light pulses of high peak intensity leading to material ablation. Thermal response is normally produced by CW or long-pulse laser radiation, when larger power delivered to organs is converted into heat and destroys the surrounding tissue. How tissues react to laser radiation in particular depends on the chosen wavelength, mode of operation, pulse duration, and energy, as well as the laser spot size.

5. 5. 3 Laser in Industry

Industry is another important field for implementing laser. Here we just provide some aspects of its applications.

(1) Laser cutting

Laser cutting is a technology that uses a laser to cut materials, and is typically used for industrial manufacturing applications, but is also starting to be used by schools, small businesses and hobbyists. Laser cutting works by directing the output of a high-power laser, by computer, at the material to be cut. The material then either melts, burns, vaporizes away, or is blown away by a jet of gas, leaving an edge with a high-quality surface finish.

Its advantages include easier work holding and reduced contamination of workpiece. Precision may be better since the laser beam does not wear during the process. There is also a reduced chance of warping the material that is being cut, as laser systems have a small heat-affected zone.

(2) Laser peening

Laser peening, or laser shock peening (LSP), is a process of hardening or peening metal using a powerful laser. Laser peening can impart a layer of residual compressive stress on a surface that is four times deeper than attainable from conventional shot peening treatments. A coating, usually black tape or paint, is applied to absorb the energy. Short energy pulses are then focused to explode the ablative coating, producing a shock wave. The beam is then repositioned and the process is repeated, creating an array of slight indents of compression and depth with about 5% ~7% cold work. A translucent layer, usually consisting of water, is required over the coating and acts as a tamp, directing the shock wave into the treated material. This computer-controlled process is then repeated, often as many as three times, until the desired compression level is reached, producing a compressive layer as deep as 1~2 mm average.

Laser peening is often used to improve the fatigue resistance of highly stressed critical turbine engine components, and the laser (or component) is typically manipulated by an industrial robot.

(3) Laser graving

Laser graving, or laser marking, is the practice of using lasers to engrave or mark an object. The technique does not involve the use of inks, nor does it involve tool bits which contact the engraving surface and wear out. These properties distinguish laser engraving from alternative engraving or marking technologies where bit heads have to be replaced regularly or inks have to be used.

The impact of laser engraving has been more pronounced for specially-designed "laserable" materials. These include laser-sensitive polymers and novel metal alloys.

The term laser marking is also used as a generic term covering a broad spectrum of surfacing techniques including printing, hot-branding and laser bonding. The machines for laser engraving and laser marking are the same, so that the two terms are usually exchangeable.

(4) Laser bonding

Laser bonding is a marking technique that uses lasers and other forms of radiant energy to bond an additive marking substance to a wide range of substrates.

Invented in 1997, this patent protected technology delivers permanent marks on metals, glass and ceramic parts for a diverse range of industrial and manual applications, ranging from aerospace to the awards & engraving industries. It differs from the more widely known techniques of laser engraving and laser ablation in that it is an additive process, adding material to the substrate to form the marking instead of removing it as in those techniques.

5.5.4 Laser Cooling

A technique that has recent success is laser cooling. This involves atom trapping, a method where a number of atoms are confined in a specially shaped arrangement of electric and magnetic fields. Shining particular wavelengths of laser light at the ions or atoms slows them down, thus cooling them. As this process is continued, they are all slowed and have the same energy level, forming an unusual arrangement of matter known as a Bose-Einstein condensate.

Laser cooling refers to the number of techniques in which atomic and molecular samples are cooled through the interaction with one or more laser light fields. The first example of laser cooling, and also still the most common method of laser cooling (so much so that it is still often referred to as "laser cooling") is Doppler cooling.

(1) Doppler cooling

Doppler cooling, which is usually accompanied by a magnetic trapping force to give a magneto-optical trap, is by far the most common method of laser cooling. It is used to cool low density gasses down to the Doppler cooling limit, which for Rubidium 85 is around 150 microkelvin. As Doppler cooling requires a very particular energy level structure, known as a closed optical loop, the method is limited to a small handful of elements.

In Doppler cooling, the frequency of light is tuned slightly below an electronic transition in the atom. Because the light is detuned to the "red" (i. e. at lower frequency) of the transition, the atoms will absorb more photons if they move towards the light source, due to the Doppler effect. Thus if one applies light from two opposite directions, the atoms will always scatter more photons from the laser beam pointing opposite to their direction of motion. In each scattering event the atom loses a momentum equal to the momentum of the photon. If the atom, which is now in the excited state, emits a photon spontaneously, it will be kicked by the same amount of momentum but in a random direction. The result of the absorption and emission process is to reduce the speed of the atom, provided its initial speed is larger than the recoil velocity from scattering a single photon. If the absorption and emission are repeated many times, the mean velocity, and therefore the kinetic energy of the atom will be reduced. Since the temperature of an ensemble of atoms is a measure of the random internal kinetic energy, this is equivalent to cooling the atoms.

(2) Other methods of laser cooling

Several somewhat similar processes are also referred to as laser cooling, in which photons are used to pump heat away from a material and thus cool it. The phenomenon has been demonstrated via anti-Stokes fluorescence, and both electroluminescent upconversion and photoluminescent upconversion have been studied as means to achieve the same effects. In many of these, the coherence of the laser light is not essential to the process, but lasers are typically used to achieve a high irradiance.

Laser cooling is primarily used for experiments in quantum physics to achieve temperatures of near absolute zero (−273.15 ℃, −459.67 °F). This is done to observe the unique quantum effects that can only occur at this heat level. Generally, laser cooling has been only used on the atomic level to cool down elements. This may soon change, as a new breakthrough in the technology has successfully cooled a macro-scale object to near absolute zero.

New Words and Expressions

abdominal　腹部的
ablation　消融
azimuth　方位，方位角
beam waist　束腰
binocular　双目镜，双筒镜
bleach　漂白
combat　战斗
cylindrical　圆柱形的
cytometry　血细胞计数
dazzle　使目眩，使眼花
depict　描绘
graving　雕刻
heterodyne　外差法
heuristic　启发式的，探索的
humidity　湿度、湿气
hyperbola　双曲线
lethal　致命（性）的
peening　锤击，敲打
polymer　聚合物
polynomial　多项式
retain　保持
rifle　步枪，来复枪
turbine　涡轮（机）
warp　变形

Translation

1. As we have noted, the fluorescent emission is not directional, but some of this light will travel along the cavity's axis.

翻译： 正如我们所注意到的，荧光发射是不定向的，但是其中一些光会沿着空腔的轴线传播。

2. Its coherence length is great compared with the optical length of the cavity.

翻译： 它的相干长度与光腔的光学长度相比很大。

3. If the packet returns out of phase with the wave that is still being emitted by the atom, it will interfere destructively with that wave and effectively terminate the emission.

翻译：如果返回的波束仍与原子发射的波不同步，它将破坏性地干扰该波并有效地终止发射。

4. We can, if we wish, say that such a wave has been reabsorbed, but the effect is as if the wave never existed.

翻译：如果我们愿意的话，可以说这样的波已经被重新吸收了，但结果就好像这个波从未存在过一样。

5. Naturally, the ray will not describe the precise field distribution because of the effects of diffraction.

翻译：自然地，由于衍射的影响，射线不能描述精确的场分布。

6. In our earlier treatment of diffraction, we assumed that a uniform wavefront passed through the diffracting screen.

翻译：在之前的衍射处理中，我们假设通过衍射屏的波前是均匀的。

7. The beam propagates, both inside and outside the cavity, in such a way that it retains its Gaussian profile.

翻译：光束在腔体内外传播时保持高斯分布。

8. To find the size and location of the beam waist, we argue that the radius of curvature of the beam wavefront at the positions of the mirrors must be exactly equal to the radii of the mirrors themselves.

翻译：为了求出光束的大小和位置，我们认为在反射镜位置的光束波前曲率半径必须完全等于反射镜本身的半径。

9. Both radius and spot size must be matched to ensure effective mode matching.

翻译：半径和光斑大小必须匹配，以确保有效的模式匹配。

10. The most commonly used laser resonators are composed of two spherical or flat mirrors facing each other, as shown.

翻译：如图所示，最常用的激光谐振器是由两个相互面对的球面或平面镜组成的。

11. A special case of a symmetrical configuration is the concentric resonator that consists of two mirrors separated by twice their radius.

翻译：对称结构的一个特殊情况是同心谐振器，它由两个反射镜组成，两个反射镜的距离是其半径的两倍。

12. The power or energy from an oscillator with specific spatial, temporal, or spectral properties can be increased by adding one or more amplifying stages to the laser system.

翻译：通过在激光系统中增加一个或多个放大级，可以提高具有特定空间、时间或光谱特性的振荡器的功率或能量。

Unit 6
Optical Fiber Communication

6.1 The Development of Optical Communication Systems

The first generation of lightwave transmission systems utilized GaAs lasers operating at a wavelength of about 0.85 μm. Relatively high values of the loss and dispersion coefficients restricted the repeater spacing to ~10 km and the bit rate to ~100 Mb/s.

The second generation of systems made use of the wavelength region around 1.3 μm, where fiber dispersion is negligible. The use of InGaAs lasers coupled with the relatively low fiber loss allowed a repeater spacing of about 20 km. However, the bit rate had to be bellow 100 Mb/s because of modal dispersion in multimode fibers. This problem was overcome with the use of single-mode fibers; the absence of chromatic dispersion near 1.3 μm then allowed much higher bit rates (up to 2 Gb/s). However, the repeater spacing (~50 km) was limited by the fiber loss at this wavelength.

Minimum fiber loss occurs around 1.55 μm. The third generation of optical communication systems is therefore based on 1.55 μm InGaAs lasers. At this wavelength the repeater spacing can easily exceed 100 km for moderate bit rates. At high bit rates ($B>$ 1 Gb/s) the repeater spacing is limited not by the fiber loss but rather by the extent of fiber dispersion. Two distinct routes are being followed to overcome this problem. In one approach the zero-dispersion wavelength, which is about 1.3 μm for conventional silica fibers, is shifted towards the desirable 1.55 pm region by modifying the fiber characteristics. In the other approach the effect of fiber dispersion is minimized by reducing the spectral width of the 1.55 μm InGaAs laser source.

The fourth-generation of optical communication systems makes use of homodyne or heterodyne detection techniques. Because of the phase-sensitive nature of such systems, they are referred to as coherent communication system. For the same reason, they require tunable semiconductor lasers with a narrow linewidth (usually $<$10 MHz or 10^{-4} nm). Special multisection lasers have been developed to meet the demands of fourth-generation lightwave systems.

The fifth generation of lightwave systems employs erbium-doped fiber amplifiers for amplifying the transmitted optical signal periodically to compensate for fiber loss. The transmission distance can exceed a few thousand kilometers for such systems as long as fiber dispersion does not limit it. A novel approach makes use of optical solitons that use fiber nonlinearity for compensating fiber dispersion. Such solitons can travel along the fiber for thousands of kilometers without experiencing temporal broadening as long as fiber loss is compensated through periodic amplification. The development of such fifth-generation lightwave systems requires new kinds of semiconductor lasers. The use of optical solitons requires semiconductor lasers capable of producing ultrashort optical pulses (pulse width< 50 ps) at high repetition rates. Advances in the semiconductor-laser technology play an important role in realizing such high-performance lightwave systems.

6.2 Optical Fiber Characteristics

In its simplest form, an optical fiber consists of a central core surrounded by a cladding layer whose refractive index is slightly lower than that of the core. Such fibers are generally referred to as step-index fibers to distinguish them from graded-index fibers in which the refractive index of the core decreases gradually from center to the core boundary. Figure 6-1 shows schematically the cross section and the refractive-index profile of a step-index fiber. Two parameters which characterize the fiber are the relative core-cladding index difference Δ defined by

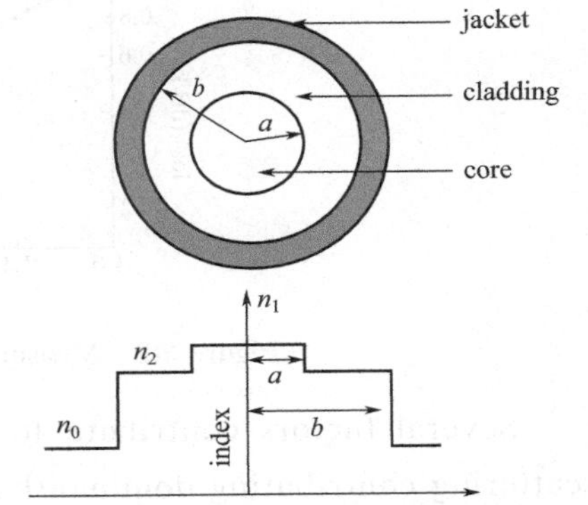

Figure 6-1 Schematic illustration of the cross section and the refraction-index profile of a step-index fiber

$$\Delta = \frac{n_1 - n_2}{n_1} \tag{6.1}$$

and the normalized frequency V defined by

$$V = k_0 a (n_1^2 - n_2^2)^{1/2} \tag{6.2}$$

where $k_0 = 2\pi/\lambda$, a is the core radius, and λ is the wavelength of light.

The parameter V determines the number of mode supported by the fiber. A step-index fiber supports a single mode if $V < 2.405$. The fibers designed to satisfy this condition are called single-mode fibers. The main difference between the single-mode and multimode fibers is the core size. The core radius $a = 25 \sim 30$ μm for typical multimode fibers. However, single-mode fibers with a typical value of $\Delta \approx 3 \times 10^{-3}$ require a to be in the range $2 \sim 4$ μm. The numerical value of the outer radius b is less critical as long as it is large enough to confine the fiber modes entirely. Typically $b = 50 \sim 60$ μm for both single-mode and multimode fibers.

6.2.1 Optical Losses

An important fiber parameter is a measure of power loss during inside the fiber. If P_0 is the power launched at the input of a fiber of length L, the transmitted power P_T is given by

$$P_T = P_0 \exp(-aL) \tag{6.3}$$

where a is the attenuation constant, commonly referred to as the express of the fiber loss in units of dB/km by using the relation

$$\alpha_{dB} = \frac{10}{L} \lg\left(\frac{P_T}{P_0}\right) = 4.343\alpha \tag{6.4}$$

where Eq. (6.4) is used to relate α_{dB} and α.

The fiber loss depends on the wavelength of light. Figure 6-2 shows the loss spectrum of a silica fiber. This fiber exhibits a minimum loss of about 0.2 dB/km near 1.55 μm. The loss is considerably higher at shorter wavelengths, reaching a level of 1~10 dB/km in the visible region. Note, however, that even a 10 dB/km loss corresponds to an attenuation constant of only $a \approx 2\times10^{-5}\text{cm}^{-1}$.

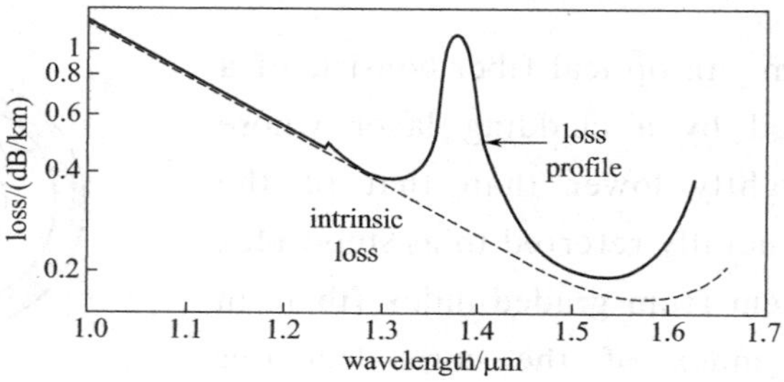

Figure 6-2 Measured loss spectrum of a single-mode silica fiber

Several factors contribute to the loss spectrum, material absorption and Rayleigh scattering contributing dominantly. Other factors that may contribute to the fiber loss are bending losses and boundary losses (due to scattering at the core-cladding boundary). The total loss of a fiber link in the optical communication systems also includes the splice losses that occur when two fiber pieces are joined together. Advances in fiber technology have reduced splice losses to a level of ~0.01 dB/km.

6.2.2 Chromatic Dispersion

When an electromagnetic wave interacts with bound electrons of a dielectric, the medium response in general depends on the optical frequency ω. This property, referred to as chromatic dispersion, manifests through the frequency dependence of the refractive index $n(\omega)$. On a fundamental level, the origin of chromatic dispersion is related to the characteristic resonance frequencies at which the medium absorbs the electromagnetic radiation through oscillations of bound electrons. Far from the medium resonances, the refractive index is well approximated to the Sellmeier equation

$$n^2(\omega) = 1 + \sum_{j=1}^{m} \frac{B_j\omega_j^2}{\omega_j^2 - \omega^2} \tag{6.5}$$

where ω_j is the resonance frequency and B_j is the strength of j_{th} resonance. The sum in Eq. (6.5) extends over all material resonances that contribute to the frequency range of interest. In the case of optical fibers, the parameters B_j and ω_j, are obtained experimentally by fitting the measured dispersion curve to Eq. (6.5) with $m = 3$ and depend on the core constituent.

Fiber dispersion plays a critical role in propagation of short optical pulses since different spectral components are associated with the pulse travel at different speeds given by $c/n(\omega)$. Even when nonlinear effects are not important, dispersion-induced pulse broadening can be detrimental for optical communication systems.

Mathematically, the effects of fiber dispersion are accounted for by expanding the mode-propagation constant β in a Taylor series about the frequency ω_0 at which the pulse spectrum is centered

$$\beta(\omega) = n(\omega)\frac{\omega}{c} = \beta_0 + \beta_1(\omega - \omega_0) + \frac{1}{2}\beta_2(\omega - \omega_0)^2 + \cdots \tag{6.6}$$

where

$$\beta_m = \left.\frac{\mathrm{d}^m \beta}{\mathrm{d}\omega^m}\right|_{\omega = \omega_0} \quad (m = 0, 1, 2, \cdots) \tag{6.7}$$

The parameters β_1 and β_2 are related to the refractive index n and its derivatives through the relations

$$\beta_1 = \frac{1}{v_g} = \frac{n_g}{c} = \frac{1}{c}\left(n + \omega\frac{\mathrm{d}n}{\mathrm{d}\omega}\right) \tag{6.8}$$

$$\beta_2 = \frac{1}{c}\left(2\frac{\mathrm{d}n}{\mathrm{d}\omega} + \omega\frac{\mathrm{d}^2 n}{\mathrm{d}\omega^2}\right) \tag{6.9}$$

where n_g is the group index and v_g is the group velocity. Physically speaking, the envelope of an optical pulse moves at the group velocity while the parameter β_2 represents dispersion of the group velocity and is responsible for pulse broadening. This phenomenon is known as the group-velocity dispersion (GVD), and β_2 is the GVD parameter. The coefficient β_3 appearing in that term is called the third-order dispersion (TOD) parameter. Such higher-order dispersive effects can distort ultrashort optical pulses both in the linear and nonlinear regimes.

Figure 6-3 and Figure 6-4 show how n, n_g, and β_2 vary with wavelength λ in fused silica. The most notable feature is that β_2 vanishes at a wavelength of about 1.3 μm and becomes negative for longer wavelengths. This wavelength is referred to as the zero-dispersion wavelength and is denoted as λ_D.

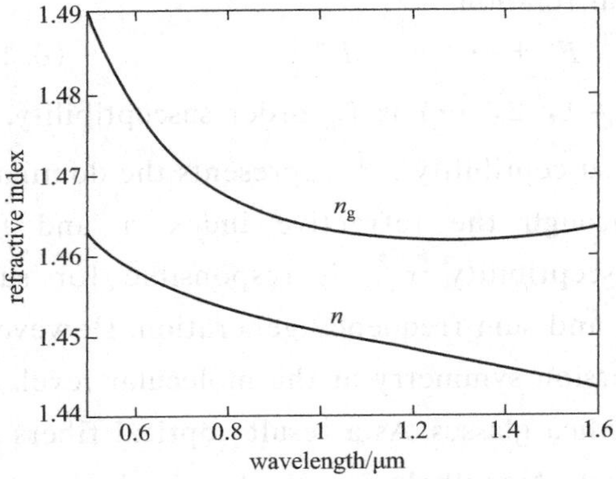

Figure 6-3 Variation of refractive index n and group index n_g with wavelength for fused silica

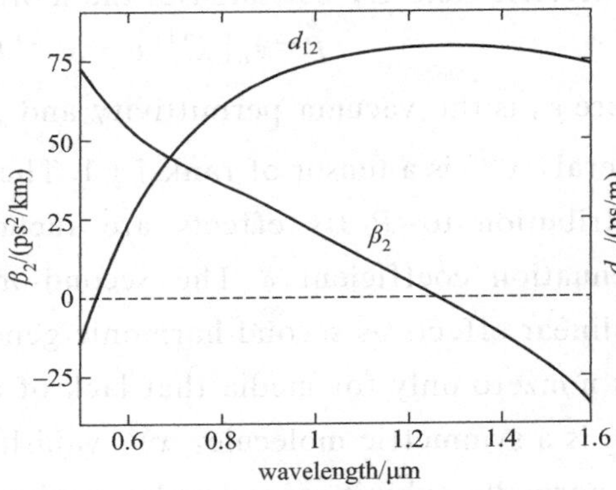

Figure 6-4 Variation of β_2 with wavelength for fused silica

It is possible to design dispersion-flattened optical fibers having low dispersion over a relatively large wavelength range 1.3～1.6 μm. This can be achieved by using multiple cladding layers.

Nonlinear effects in optical fibers can manifest qualitatively different behaviors depending on the sign of the GVD parameter. For wavelengths such that $\lambda = \lambda_D$, the fiber is said to exhibit normal dispersion as $\beta_2 > 0$. In the normal-dispersion regime, high-frequency (blue-shifted) components of an optical pulse travel slower than low-frequency (red-shifted) components of the same pulse. By contrast, the opposite occurs in the anomalous dispersion regime in which $\beta_2 < 0$. As seen in Figure 6-4, silica fibers exhibit anomalous dispersion when the light wavelength exceeds the zero-dispersion wavelength ($\lambda = \lambda_D$). The anomalous-dispersion regime is of considerable interest for the study of nonlinear effects because it is in this regime that optical fibers support solitons through a balance between the dispersive and nonlinear effects.

An important feature of chromatic dispersion is that pulses at different wavelengths propagate at different speeds inside the fiber because of the group-velocity mismatch. This feature leads to a walk-off effect that plays an important role in the description of the nonlinear phenomena involving two or more overlapping optical pulses. More specifically, the nonlinear interaction between two optical pulses ceases to occur when the faster moving pulse has completely walked through the slower moving pulse. The group-velocity mismatch plays an important role in the case of the nonlinear effects involving cross-phase modulation.

6.2.3 Fiber Nonlinearities

The response of any dielectric to light becomes nonlinear for intense electromagnetic fields, and optical fibers are no exception. On a fundamental level, the origin of nonlinear response is related to a harmonic motion of bound electrons under the influence of an applied field. As a result, the total polarization P induced by electric dipoles is not linear in the electric field E, but satisfies the more general relation

$$P = \varepsilon_0 [x^{(1)} E + x^{(2)} E^2 + x^{(3)} E^3 + \cdots + x^{(n)} E^n] \tag{6.10}$$

where ε_0 is the vacuum permittivity and $x^{(j)}$ ($j = 1, 2, \cdots$) is j_{th} order susceptibility. In general, $x^{(j)}$ is a tensor of rank $j+1$. The linear susceptibility $x^{(j)}$ represents the dominant contribution to P. Its effects are included through the refractive index n and the attenuation coefficient α. The second-order susceptibility $x^{(2)}$ is responsible for such nonlinear effects as second-harmonic generation and sum-frequency generation. However, it is nonzero only for media that lack of an inversion symmetry at the molecular level. As SiO_2 is a symmetric molecule, $x^{(2)}$ vanishes for silica glasses. As a result, optical fibers do not normally exhibit second-order nonlinear effects. Nonetheless, defects or color centers inside the fiber core can contribute to second-harmonic generation under certain conditions.

The lowest-order nonlinear effects in optical fibers originate from the third-order susceptibility $x^{(3)}$, which is responsible for phenomena such as third-harmonic generation, four-wave mixing, and nonlinear refraction. However, unless special efforts are made to achieve phase matching, the nonlinear processes which involve the generation of new frequencies are not efficient in optical fibers. Most of the nonlinear effects in optical fibers therefore originate from nonlinear refraction, a phenomenon that refers to the intensity dependence of the refractive index resulting from the contribution of $x^{(3)}$. The intensity dependence of the refractive index leads to a large number of interesting nonlinear effects; the two most widely studied are self-phase modulation (SPM) and cross-phase modulation (XPM). SPM refers to the self-induced phase shift experienced by an optical field during its propagation in optical fibers. SPM is responsible for spectral broadening of ultrashort pluses and the existence of optical solitons in the anomalous-dispersion regime of fiber. XPM refers to the nonlinear phase shift of an optical field induced by a copropagating field at a different wavelength. XPM is responsible for asymmetric spectral broadening of copropagating optical pulses.

The nonlinear effects governed by the third-order susceptibility $x^{(3)}$ are elastic in the sense that no energy is exchanged between the electromagnetic field and the dielectric medium. A second class of nonlinear effects results from stimulated inelastic scattering in which the optical field transfers part of its energy to the nonlinear medium. Two important nonlinear effects in optical fibers fall in this category; both of them are related to vibrational excitation modes of silica. These phenomena, known as stimulated Raman scattering (SRS) and stimulated Brillouin scattering (SBS), are among the first nonlinear effects studied in optical fibers. The main difference between the two is that optical photons participate in SRS while acoustic phonons participate in SBS.

In a simple quantum-mechanical picture applicable to both SRS and SBS, a photon of the incident field (called the pump) is annihilated to create a photon at a lower frequency (belonging to the Stokes wave) and a phonon with the right energy and momentum to conserve the energy and the momentum. Of course, a higher-energy photon at the so-called anti-Stokes frequency can also be created if a phonon of right energy and momentum is available. Even though SRS and SBS are very similar in their origin, different dispersion relations for acoustic and optical phonons lead to some basic differences between the two. A fundamental difference is that SBS in optical fibers occurs only in the backward direction whereas SRS can occur in both directions.

6.3 The Propagation of Optical Beam in Fibers

For an understanding of the nonlinear phenomena in optical fibers, it is necessary to consider the theory of electromagnetic wave propagation in dispersive nonlinear media.

Maxwell's equations can be used to obtain the wave equation that describes light propagation in optical fibers, i. e.,

$$\nabla^2 \widetilde{E}(r,\omega) + \varepsilon(\omega) k_0^2 \widetilde{E}(r,\omega) = 0 \tag{6.11}$$

where $k_0=\omega/c$, $\widetilde{E}(r,\omega)$ is the Fourier transform of $E(r,t)$ defined as

$$\widetilde{E}(r,\omega)=\int_{-\infty}^{+\infty}E(r,t)\exp(i\omega t)\mathrm{d}t \tag{6.12}$$

and the dielectric constant $\varepsilon(\omega)$ can be written as

$$\varepsilon(\omega)=1+\widetilde{x}^{(1)}(\omega)+\varepsilon_{\mathrm{NL}} \tag{6.13}$$

$$\varepsilon_{\mathrm{NL}}=\frac{3}{4}\widetilde{x}^{(3)}(\omega)\ |E(r,t)|^2 \tag{6.14}$$

According to Eq. (6.11), the theory of pulse propagation in nonlinear dispersive media in the slowly varying envelope approximation with the assumption that the spectral width of the pulse is much smaller than the frequency of the incident radiation can be obtained.

6.3.1 Mode Characteristics

At any frequency ω, optical fibers can support a finite number of guided modes whose spatial distributions $\widetilde{E}(r,\omega)$ are a solution of the wave Eq. (6.11) and satisfy all appropriate boundary conditions. In addition, the fiber can support a continuum of unguided radiation modes. Although the inclusion of radiation modes is crucial in problems involving transfer of power between bounded and radiation modes, they do not play an important role in the discussion of nonlinear effects.

Because of the cylindrical symmetry of fibers, it is useful to express the wave equation in cylindrical coordinates. The wave equation for $\widetilde{E}_z$ is easily solved by using the method of separation of variables, resulting in the following general form

$$\widetilde{E}(r,\omega)=A(\omega)F(\rho)\exp(\pm im\Phi)\exp(iBz) \tag{6.15}$$

where A is a normalization constant, β is the propagation constant, m is an integer, and

$$F(\rho)=J_m(K\rho),\rho\leqslant\alpha \tag{6.16}$$

$$F(\rho)=K_m(\gamma\rho),\rho\leqslant\alpha \tag{6.17}$$

In Eq. (6.16) and Eq. (6.17), J_m is the Bessel function, K_m is the modified Bessel function, α is the core radius and

$$K=(n_1^2k_0^2-B^2)^{1/2} \tag{6.18}$$

$$\gamma=(B^2-n_2^2k_0^2)^{1/2} \tag{6.19}$$

where n_1 and n_2 are the refractive index of fiber core and cladding layer, respectively.

The derivation of the eigenvalue equation whose solutions determine the propagation constant β for the fiber modes is left to the references, here we write the eigenvalue equation directly, i. e.

$$\left[\frac{J'_m(K\alpha)}{KJ_m(K\alpha)}+\frac{K'_m(\gamma\alpha)}{\gamma k_m(\gamma\alpha)}\right]\left[\frac{J'_m(K\alpha)}{KJ_m(K\alpha)}+\frac{n_2^2}{n_1^2}\frac{K'_m(\gamma\alpha)}{\gamma k_m(\gamma\alpha)}\right]=\frac{mBk_0(n_1^2-n_2^2)}{\alpha n_1K^2\gamma^2} \tag{6.20}$$

where a prime denotes differentiation with respect to the argument.

The eigenvalue equation [Eq. (6.20)] in general has several solutions for β for each integer value of m. It is customary to express these solutions by β_{mn}, where both m and n take integer values. Each eigenvalue β_{mn} corresponds to one specific mode supported by the fiber. The corresponding modal field distribution is obtained from Eq. (6.15). It turns out that there are two types of fiber modes, designated as HE_{mn} and EH_{mn}. For $m=0$, these modes are analogous to the transverse electric (TE) and transverse magnetic (TM) modes of a planar waveguide because the axial component of the electric field, or the magnetic field, vanishes. However, for $m>0$, fiber modes become hybrid, i. e., all six components of the electromagnetic field are nonzero.

6.3.2 Optical Pulse Propagation and Pulse Spreading in Fibers

The study of most nonlinear effects in optical fibers involves the use of short pulses with widths ranging from ~10 ns to 10 fs. When such optical pulses propagate inside a fiber, both dispersive and nonlinear effects influence their shape and spectrum.

A basic equation that governs propagation of optical pulses in nonlinear dispersive fibers can be written as

$$\nabla^2 E - \frac{1}{C^2}\frac{\partial^2 E}{\partial t^2} = \mu_0 \frac{\partial^2 P_L}{\partial t^2} + \mu_0 \frac{\partial^2 P_{NL}}{\partial t^2} \tag{6.21}$$

It is necessary to make several simplifying assumptions before solving Eq. (6.21). First, P_{NL} is treated as a small perturbation to P_L. This is justified because nonlinear changes in the refractive index are $<10^{-6}$ in practice. Second, the optical field is assumed to maintain its polarization along the fiber length so that a scalar approach is valid. This is not really the case, unless polarization-maintaining fibers are used, but the approximation works quite well in practice. Third, the optical field is assumed to be quasi-monochromatic, i. e., the pulse spectrum, centered at ω_0, is assumed to have a spectral width $\Delta\omega$ such that $\Delta\omega/\omega_0 \ll 1$. Since $\omega_0 \approx 10^{15}\,s^{-1}$, the last assumption is valid for pulses as short as 0.1 ps.

Eq. (6.11) can be solved by using the method of separation of variables. If we assume a solution of the form

$$\tilde{E}(r,\omega-\omega_0) = F(x,y)\,\tilde{A}(z,\omega-\omega_0)\exp(i\beta_0 z) \tag{6.22}$$

where $\tilde{A}$ is a slowly varying function of z and β_0 is the wave number. Substituting from Eq. (6.22) into Eq. (6.21), we obtain the following two equations

$$\frac{\partial^2 F}{\partial x^2} + \frac{\partial^2 F}{\partial y^2} + [\varepsilon(\omega)k_0^2 - \tilde{\beta}^2]F = 0 \tag{6.23}$$

$$2i\beta_0 \frac{\partial \tilde{A}}{\partial z} + (\tilde{\beta}^2 - \beta_0^2)\tilde{A} = 0 \tag{6.24}$$

where

$$\varepsilon = (n+\Delta n)^2 \approx n^2 + 2n\Delta n \tag{6.25}$$

$$\tilde{\beta} = \beta(\omega) + \Delta\beta \tag{6.26}$$

$$\Delta n = n_2 |E|^2 + \frac{i\tilde{\alpha}}{2k_0} \tag{6.27}$$

$$\Delta\beta = \frac{k_0 \iint_{-\infty}^{+\infty} \Delta n\,|F(x,y)|^2\,dx\,dy}{\iint_{-\infty}^{+\infty} |F(x,y)|^2\,dx\,dy} \tag{6.28}$$

Eq. (6.23) can be solved using first-order perturbation theory. We replace ε with n_2 and obtain the modal distribution $F(x, y)$, and the corresponding wave number $\beta(\omega)$. For a single-mode fiber, $F(x, y)$ corresponds to the modal distribution of the fundamental fiber mode HE_{11}, or by the Gaussian approximation. Eq. (6.24) can be further expressed as

$$\frac{\partial A}{\partial z}+\beta_1\frac{\partial A}{\partial z}+\frac{i}{2}\beta_2\frac{\partial^2 A}{\partial t^2}+\frac{\alpha}{2}A=i\gamma|A|^2A \tag{6.29}$$

Eq. (6.29) describes propagation of picosecond optical pulse in single-mode fibers. It is often referred to as the nonlinear Schrödinger (NLS) equation because it can be reduced to that form under certain conditions. It includes the effects of fiber losses through α, of chromatic dispersion through β_1 and β_2, and of fiber nonlinearity through γ.

The combined effects of group-velocity dispersion (GVD) and self-phase modulation (SPM) on optical pulses propagating inside a fiber can be studied by solving a pulse-propagation equation.

6.3.3 Dispersion Management

In a fiber-optic communication system, information is transmitted over a fiber by using a coded sequence of optical pulses whose width is determined by the bit rate B of the system. Dispersion-induced broadening of pulses is undesirable as it interferes with the detection process and leads to errors if the pulse spreads outside its allocated bit slot ($T_B = 1/B$). Clearly, GVD limits the bit rate B for a fixed transmission distance L. The dispersion problem becomes quite serious when optical amplifiers are used to compensate for fiber losses because L can exceed thousands of kilometers for long-haul systems.

Even though operation at the zero-dispersion wavelength is most desirable from the standpoint of pulse broadening, other considerations may preclude such a design. For example, at most one channel can be located at the zero-dispersion wavelength in a wavelength-division-multiplexed (WDM) system. Moreover, strong four-wave mixing occurring when GVD is relatively low forces WDM systems to operate away from the zero-dispersion wavelength so that each channel has a finite value of β_2. Of course, GVD-induced pulse broadening then becomes serious concern. The technique of dispersion management provides a solution to this dilemma. It consists of combining fibers with different characteristics such that the average GVD of the entire fiber link is quite low while the GVD of each fiber section is chosen to be large enough to make the four-wave-mixing effects negligible. In practice, a periodic dispersion map is used with a period equal to the amplifier spacing (typically 50～100 km). Amplifiers compensate for accumulated fiber losses in each section. Between each pair of amplifiers, just two kinds of fibers, with opposite signs of β_2, are combined to reduce the average dispersion to a small value. When the average GVD is set to zero, dispersion is totally compensated.

When the bit rate of a single channel exceeds 100 Gb/s, one must use ultrashort pulses (width～1 ps) in each bit slot. For such short optical pulses, the pulse spectrum becomes broad enough that it is difficult to compensate GVD over the entire bandwidth of the pulse (because of the frequency dependence of β_2). The simplest solution to this problem is provided by fibers, or other devices, designed such that both β_2 and β_3 are compensated simultaneously. For a fiber link containing two different fibers of lengths L_1 and L_2, the conditions for broadband dispersion compensation are given by

$$\beta_{21}L_1+\beta_{22}L_2=0 \quad \text{and} \quad \beta_{31}L_1+\beta_{32}L_2=0 \tag{6. 30}$$

where β_{2j} and β_{3j} are the GVD and TOD (third-order dispersion) parameters for fiber of length L_1 $(j=1, 2)$.

It is generally difficult to satisfy both conditions simultaneously over a wide wavelength range. However, for a 1 ps pulse, it is sufficient to satisfy Eq. (6. 30) over a 4~5 nm bandwidth. This requirement is easily met for dispersion-compensating fibers (DCFs), especially designed with negative values of β_3 (sometimes called reverse-dispersion fibers). Fiber gratings, liquid-crystal modulators, and other devices can also be used for this purpose.

6. 3. 4 Solitons

A fascinating manifestation of the fiber nonlinearity occurs through optical solitons, formed as a result of the interplay between the dispersive and nonlinear effects. The word "soliton" refers to special kinds of wave packets that can propagate undistorted over long distances. Solitons have been discovered in many branches of physics. In the context of optical fibers, not only are solitons of fundamental interest but they have also found practical applications in the field of fiber-optic communications. The use of solitons for optical communications was first suggested in 1973. By the year 1999, several field trials making use of fiber solitons have been completed.

The first-order soliton corresponds to the case of a single eigenvalue of nonlinear wave equations. It is referred to as the fundamental soliton because its shape does not change on propagation. Fundamental solitons can form in optical fibers at power levels available from semiconductor lasers even at a relatively high bit rate of 20 Gb/s.

Starting in 1988, most of the experimental work on fiber solitons was devoted to their applications in fiber-optic communication systems. Such systems make use of fundamental solitons for representing "1" bits in a digital bit stream. In a practical situation, solitons can be subjected to many types of perturbations as they propagate inside an optical fiber. Examples of perturbations include fiber losses, amplifier noise (if amplifiers are used to compensate fiber losses), third order dispersion, and intrapulse Raman scattering.

6. 4 The Impacts of Fiber Nonlinearities

As seen in Section 6. 2. 3, the nonlinear effects occurring inside optical fibers limit the maximum power levels and degrade the communication quality, including stimulated Brillouin scattering (SBS), stimulated Raman scattering (SRS), self-phase modulation

(SPM) and cross-phase modulation (XPM), and four-wave mixing (FWM).

6.4.1 Stimulated Brillouin Scattering

SBS in optical fibers was first observed in 1972 and has been studied extensively since then because of its implications for lightwave systems. SBS generates a Stokes wave propagating in the backward direction. The frequency of the Stokes wave is downshifted by an amount that depends on the wavelength of incident signal. This shift is known as the Brillouin shift and is about 11 GHz in the wavelength region near 1.55 μm. The intensity of the Stokes wave grows exponentially once the input power exceeds a threshold value.

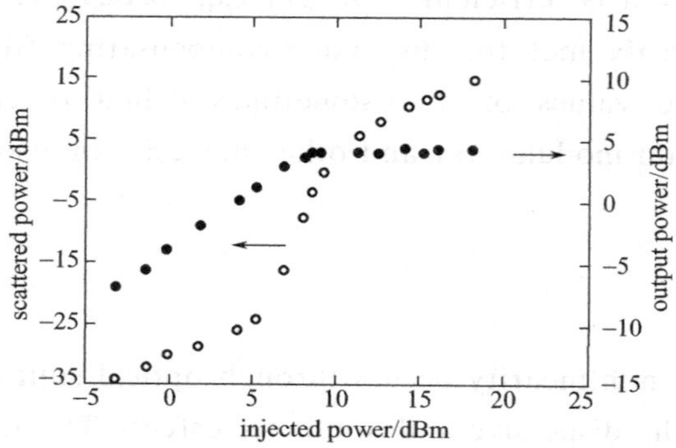

Figure 6-5 Output signal power (solid circles) and reflected SBS power (empty circles) as a function of power injected into a 13 km-long fiber

Figure 6-5 shows variations in the transmitted reflected power (through SBS) for a 13 km-long dispersion-shifted fiber as the injected CW power is increased from 0.5 mW to 50 mW. No more than 3 mW could be transmitted through the fiber in this experiment after the onset of SBS.

In lightwave systems, the optical signal is in the form of a time-dependent signal composed of an arbitrary sequence of 1 and 0 bits. One would expect the Brillouin threshold of such a signal to be higher than that of a CW beam. Considerable attention has been paid to estimating the Brillouin threshold and quantifying the SBS limitations for practical lightwave systems. The amount by which the threshold power increases depends on the modulation format used for data transmission. In the case of a coherent transmission scheme, the SBS threshold also depends on whether the amplitude, phase, or frequency of the optical carrier modulated for information coding. Most lightwave systems modulate amplitude of the optical carrier and use the so-called on-off keying scheme.

In modern wavelength-division-multiplexed (WDM) systems, fiber losses are compensated periodically using optical amplifiers. An important question is how amplifiers affect the SBS process. If the Stokes wave were amplified by amplifiers, it would accumulate over the entire link and grow enormously. Fortunately, periodically amplified lightwave systems typically employ an optical isolator within each amplifier that blocks the passage of the Stokes wave. However, the SBS growth between two amplifiers is still undesirable for two reasons. First, it removes power from the signal once the signal power exceeds the threshold level. Second, it induces large fluctuations in the remaining signal, resulting in degradation of both the SNR and the bit-error rate (BER). For these reasons, single-channel powers are invariably kept below the SBS threshold and are limited in practice to below 10 mW.

Fiber gratings can also be used to increase the SBS threshold. The Bragg grating is designed such that it is transparent to the forward-propagating pump beam, but the spectrum of the Stokes wave generated through SBS falls entirely within its stop band. A single grating, placed suitably in the middle, may be sufficient for relatively short fibers. Multiple gratings need to be used for long fibers. Another approach makes one try to minimize the overlap between the optical and acoustic modes through suitable dopants.

6. 4. 2 Stimulated Raman Scattering

SRS differs from SBS in several ways. First, it generates a forward-propagating Stokes wave. Second, the Raman shift by which the frequency of the Stokes wave is down-shifted is close to 13 THz. Third, the Raman-gain spectrum is extremely broad and extends over a frequency range wider than 20 THz. Fourth, the peak value of the Raman gain is lower by more than a factor of 100 compared with that of the Brillouin gain. SRS was first observed in optical fibers in 1972. Since then, the impact of SRS on the performance of lightwave systems has been studied extensively.

The Raman threshold, the power level at which the Raman process becomes stimulated and transfers most of the signal power to the Stokes wave, can be written as

$$P_{th} = 16A_{eff}/(g_R L_{eff}) \tag{6.31}$$

where we can replace L_{eff} with $1/\alpha$ for long fiber lengths used in lightwave systems. Using $g_R = 1\times10^{13}$ m/W, P_{th} is about 500 mW at wavelengths near 1. 55 μm. Since input powers are limited to below 10 mW because of SBS, SRS is not of concern for single-channel lightwave systems.

The situation is quite different for WDM systems, which transmit simultaneously multiple channels spaced 100 GHz or so apart. The fiber link in this case acts as a Raman amplifier such that longer-wavelength channels are amplified by shorter-wavelength channels as long as their wavelength difference is within the Raman-gain bandwidth. The shortest-wavelength channel is depleted most as it can pump all other channels simultaneously. Such an energy transfer among channels can be detrimental for system performance as it depends on the bit pattern, occurring only when 1 bits are present simultaneously in the two channels acting as the pump and signal channels. The signal-dependent amplification leads to power fluctuations, which add to receiver noise and degrade the receiver performance.

The Raman crosstalk can be avoided if channel powers are made so small that Raman amplification is negligible over the fiber length. A simple model considered depletion of the highest-frequency channel in the worst case in which all channels transmit 1 bits simultaneously. A more accurate analysis should consider not only depletion of each channel but also its own amplification by shorter-wavelength channels. The effects of Raman crosstalk in a WDM system were quantified in a 1999 experiment by transmitting 32 channels, with 100 GHz spacing, over 100 km. At low input powers, SRS effects were relatively small and channel powers differed by only a few percent after 100 km. However, when the input power for each channel was increased to 3. 6 mW, the longest-wavelength channel had 70% more power than the shortest-wavelength channel. Moreover, the channel powers were distributed in an exponential fashion.

In long-haul lightwave systems, the crosstalk is also affected by the use of loss and dispersion-management schemes. Dispersion management permits high values of GVD locally while reducing it globally. Since the group-velocity mismatch among different channels is quite large in such systems, the Raman crosstalk should be reduced in a dispersion-managed system. In contrast, the use of optical amplifiers for loss management magnifies the impact of SRS-induced degradation. The reason is that inline amplifiers add broadband noise, which can seed the SRS process. As a result, noise is amplified along the link and results in degradation of the SNR. The SNR can be maintained if the channel power is increased as the number of amplifiers increases. The Raman-limited capacity of long-haul WDM systems depends on a large number of design parameters such as amplifier spacing, optical-filter bandwidth, bit rate, channel spacing, and total transmission distance.

Can Raman crosstalk be avoided by a proper system design? Clearly, reducing the channel power is the simplest solution, but it may not always be practical. An alternative scheme lets SRS occur over the whole link but cancels the Raman crosstalk by using the technique of spectral inversion. As the name suggests, if the spectrum of the WDM signal were inverted at some appropriate distance, short-wavelength channels would become long-wavelength channels and vice versa. As a result, the direction of Raman-induced power transfer will be reversed such that channel powers become nearly equal at the end of the fiber link. Complete cancelation of Raman crosstalk for a two-channel system requires spectral inversion at mid-span if GVD effects are negligible or compensated. The location of spectral inversion is not necessarily in the middle of the fiber span but changes depending on gain-loss variations. Spectral inversion can be accomplished through FWM inside a fiber to realize phase conjugation; the same technique is also useful for dispersion compensation.

6. 4. 3 Self-Phase Modulation

The intensity dependence of the refractive index leads to SPM-induced nonlinear phase shift, resulting in chirping and spectral broadening of optical pulses. Clearly, SPM can affect the performance of lightwave system. When SPM is included together with fiber dispersion and losses, the propagation of an optical bit stream through optical fibers is governed by the nonlinear Schrodinger (NLS) equation

$$\frac{\partial A}{\partial z}+\frac{i\alpha}{2}A-\frac{\beta_2}{2}\frac{\partial^2 A}{\partial T^2}+\gamma|A|^2A=0 \tag{6.32}$$

where α, β_2 and γ govern the effects of loss, GVD and SPM, respectively. All three parameters become functions of z when loss- and dispersion-management schemes are employed for long-haul lightwave systems.

It is useful to eliminate the loss term in Eq. (6. 32) with the transformation

$$A(z,T)=\sqrt{P_0}\,e^{-\alpha z/2}U(z,T) \tag{6.33}$$

where P_0 is the peak power of input pulses. Eq. (6.32) then takes the form

$$i\frac{\partial U}{\partial z}-\frac{\beta_2(z)}{2}\frac{\partial^2 U}{\partial T^2}+\gamma P_0 p(z)|U|^2U=0 \tag{6.34}$$

where power variations along a loss-managed fiber link are included through the periodic function $p(z)$. In the case of lumped amplifiers, $p(z) = e^{\alpha z}$ between two amplifiers but equals 1 at the location of each amplifier. It is not easy to solve Eq. (6.34) analytically except in some simple cases. In the specific case of $p=1$ and β_2 is constant but negative, this equation reduces to the standard NLS equation.

From a practical standpoint, the effect of SPM is to chirp the pulse and broaden its spectrum. The broadening factor can be estimated, without requiring a complete solution of Eq. (6.34), using various approximations. SPM enhances pulse broadening in the normal-GVD regime but leads to pulse compression in the anomalous-GVD regime. This behavior can be understood by noting that the SPM-induced chirp is positive in nature. As a result, the pulse goes through a contraction phase when $\beta_2<0$. This is the origin of the existence of solitons in the anomalous-GVD regime.

Another SPM-induced limitation results from the phenomenon of modulation instability occurring when the signal travels in the anomalous-GVD regime of the transmission fiber. At first sight, it may appear that modulation instability is not likely to occur for a signal in the form of a pulse train. In fact, it affects the performance of periodically amplified lightwave systems considerably. This can be understood by noting that optical pulses in a non-return-to-zero (NRZ) format system occupy the entire time slot and can be several bits long depending on the bit pattern. As a result, the situation is quasi-CW-like. As early as 1990, numerical simulations indicated that system performance of a 6000 km fiber link, operating at bit rates >1 Gb/s with 100 km amplifier spacing, would be severely affected by modulation instability if the signal propagates in the anomalous-GVD regime and is launched with peak power levels in excess of a few milliwatts.

SPM can lead to the degradation of SNR when optical amplifiers are used for loss compensation. Such amplifiers add to a signal broadband noise that extends over the entire bandwidth of amplifiers (or optical filters when they are used to reduce noise). Even close to the zero-dispersion wavelength, amplifier noise is enhanced considerably by SPM. In the case of anomalous GVD, spectral components of noise falling within the gain spectrum of modulation instability will be enhanced by this nonlinear process, resulting in further degradation of the SNR. Moreover, periodic power variations occurring in long-haul systems create a nonlinear index grating that can lead to modulation instability even in the normal-GVD regime. Both of these effects have been observed experimentally. In a 10 Gb/s system, considerable degradation in system performance was noticed after a transmission distance of only 455 km. In general, long-haul systems perform better when the average GVD of the fiber link is kept positive ($\beta_2>0$).

6.4.4 Cross-Phase Modulation

When two pulses of different wavelengths propagate simultaneously inside optical fibers, their optical phases are affected not only by SPM but also by XPM. The XPM effects are quite important for WDM lightwave systems because the phase of each optical channel depends on the bit patterns of all other channels.

The XPM effects occurring within a fiber amplifier are normally negligible because of a small length of doped fiber used. The situation changes for the L-band amplifiers, which operate in the 1570 to 1610 nm wavelength region and require fiber lengths in excess of 100 m. The effective mode area of doped fibers used in such amplifiers is relatively small, resulting in larger values of the nonlinear parameter γ and enhanced XPM-induced phase shifts. As a result, the XPM can lead to considerable power fluctuations within an L-band amplifier. A new feature is that such XPM effects are independent of the channel spacing and can occur over the entire bandwidth of the amplifier. The reason for this behavior is that all XPM effects occur before pulses walk off because of group-velocity mismatch.

The XPM effects can be reduced in modern WDM systems through the use of differential phase-shift keying (DPSK) format. The DPSK is often combined with the return-to-zero (RZ) format such that a pulse is present in every bit slot and the information is encoded only through phase variations. It is easy to understand qualitatively why XPM-induced penalties are reduced for such a lightwave system. The main reason why XPM leads to amplitude fluctuations and timing jitter when amplitude shift keying (ASK) format used is related to the random power variations that mimic the bit pattern. It is easy to see that the XPM will be totally harmless if channel powers are constant in time because all XPM-induced phase shifts are time-independent, producing no frequency and temporal shifts. Although this is not the case for an RZ-DPSK system, due to the strict periodicity, the XPM effect is greatly reduced. Physically speaking, all bits undergo nearly identical collision histories, especially if the average channel power does not vary too much along the link, resulting in negligible XPM-induced power penalties.

6.4.5 Four-Wave Mixing

Four-wave mixing (FWM) is a major source of nonlinear crosstalk for WDM lightwave systems. The physical origin of FWM-induced crosstalk, and the resulting system degradation, can be understood by noting that FWM can generate a new wave at the frequency whenever $\omega_F = \omega_i + \omega_j + \omega_k$, three waves of frequencies ω_i, ω_j, and ω_k co-propagate inside the fiber. For an M-channel system, i, j, and k vary from 1 to M, resulting in a large combination of new frequencies generated by FWM. In the case of equally spaced channels, most new frequencies coincide with the existing channel frequencies and interfere coherently with the signals in those channels. This interference depends on the bit pattern and leads to considerable fluctuations in the detected signal at the receiver. When channels are not equally spaced, most FWM components fall between the channels and add to overall noise. In both cases, system performance is affected by the loss in channel powers, but the degradation is much more severe for equally spaced channels because of the coherent nature of crosstalk.

In the case of equal channel spacing, most FWM components fall within the channel spectra and cannot be seen as clearly as in Figure 6-6 in the spectral domain.

A simple scheme for reducing the FWM-induced degradation consists of designing WDM systems with unequal channel spacings. The main impact of FWM in this case is to reduce the channel power. This power depletion results in a power penalty at the receiver whose magnitude can be controlled by varying the launched power and fiber dispersion. Experimental measurements on a WDM system, in which eight 10 Gb/s channels were transmitted over 137 km of dispersion shifted fiber, confirm the advantage of unequal channel spacings. In a 1999 experiment, this technique was used to transmit 22 channels, each operating at 10 Gb/s, over 320 km of dispersion-shifted fiber with 80 km amplifier spacing. Channel spacings ranged from 125 to 275 GHz in the 1532 to 1562 nm wavelength region and were determined using a periodic allocation scheme. The zero-dispersion wavelength of the fiber was close to 1548 nm, resulting in near phase matching of many FWM components. Nonetheless, the system performed quite well (because of unequal channel spacings) with less than 1.5 dB power penalty for all channels.

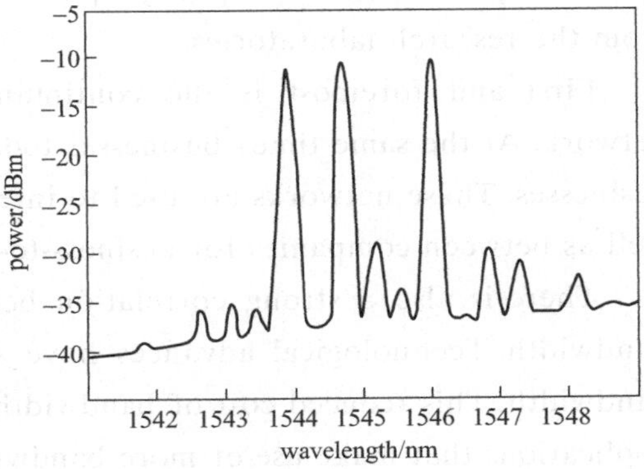

Figure 6-6 Optical spectrum measured at the output of a 25 km long fiber when three channels, each with 3 mW average power, are launched into it

The use of nonuniform channel spacings is not always practical because many WDM components require equal channel spacings. Also, this scheme is spectrally inefficient since the bandwidth of the resulting WDM signal is considerably larger compared with the case of equally spaced channels. An alternative is offered by the dispersion-management technique discussed earlier. In this case, fibers with normal and anomalous GVD are combined to form a periodic dispersion map such that GVD is locally high all along the fiber even though its average value is quite low. As a result, the FWM efficiency is negligible throughout the fiber, resulting in little FWM-induced crosstalk. As early as 1993, eight channels at 10 Gb/s could be transmitted over 280 km by using dispersion management. By 1996, the use of dispersion management had become quite common for FWM suppression in WDM systems because of its practical simplicity. FWM can also be suppressed by using fibers whose GVD varies along the fiber length. The use of chirped pulses or carrier-phase locking is also helpful for this purpose.

6.5 All Optical Networks

All optical networks are those in which either wavelength division or time division is used in new ways to form entire network structures where the messages travel in purely optical form all the way from one user location to another. There are many factors driving the development of high-capacity optical networks and their remarkably rapid transition from the research laboratories.

First and foremost is the continuing, relentless need for more capacity in the network. At the same time, businesses today rely on high-speed networks to conduct their businesses. These networks are used to interconnect multiple locations within a company as well as between companies for business-to-business transactions.

There is also a strong correlation between the increase in demand and the cost of bandwidth. Technological advances have succeeded in continuously reducing the cost of bandwidth. This reduced cost of bandwidth in turn spurs the development of a new set of applications that make use of more bandwidth and affects behavioral patterns.

Another factor causing major changes in the industry is the deregulation of the telephone industry. Traffic in networks is dominated by data as opposed to traditional voice traffic. In the past, the reverse was true, and so legacy networks were designed to efficiently support voice rather than data. Today, data transport services are pervasive and capable of producing quality of service to carry performances sensitive applications such as real-time voice and video.

Telecommunication networks have evolved during a century-long history of technological advances and social changes. Throughout the history, the digital network has evolved in three fundamental stages: asynchronous, synchronous, and optical.

The first digital networks were asynchronous networks, where each networks element's internal clock source timed its transmitting signal. Because each clock had a certain amount of variation, signals arriving and transmitting could have a large variation in timing, which often resulted in bit errors.

More importantly, as optical fiber deployment increased, no standards existed to mandate how network should format the optical signal. A myriad of proprietary methods appeared, making it difficult for network providers to interconnect equipment from different vendors.

The need for optical standards led to the creation of the synchronous optical network (SONET). SONET standardized line rates, coding schemes, bit-rates hierarchies, and operations and maintenance functionality. SONET also defined the types of network elements required, network architectures that vendors could implement, and the functionality that each node must perform. Network providers could now use different vendor's optical equipment with the confidence of at least basic interoperability.

The one aspect of SONET that has allowed it to survive during a time of tremendous changes in network capacity needs is its scalability. Base on its open-ended growth plan for higher bit rates, theoretically no upper limit exists for SONET bit rates. However, as higher bit rates are used, physical limitations in the laser sources and optical fiber begin to make the practice of endlessly increasing the bit rates on each signal an impractical solution. Additionally, connection to the networks through access rings has also increased requirements. Customers are demanding more services and options, and are carrying more and different types of data traffic. To provide full end-to-end connectivity, a new paradigm was needed to meet all the high capacity and the varied needs.

6.5.1 Components

The components used in modern optical networks include couplers, lasers, photodetectors, optical amplifiers, optical switches, filters and multiplexers.

(1) Couplers

A coupler is a versatile device and has many applications in an optical network. The simplest directional coupler is used to combine and split signals in an optical network. A 2×2 coupler consists of two input ports and two output ports, as shown in Figure 6-7. The most commonly used couplers are made by fusing two fibers together in the middle—these are called fused fiber couplers. Couplers can also be fabricated using waveguides in integrated optics. A 2×2 coupler, shown in Figure 6-7, takes a fraction of the power from input 1 and places it on output 1 and the remaining fraction $1-\alpha$ on output 2. Similarly, a fraction $1-\alpha$ of the power from input 2 is distributed to output 1 and the remaining power to output 2. We call α the coupling ratio.

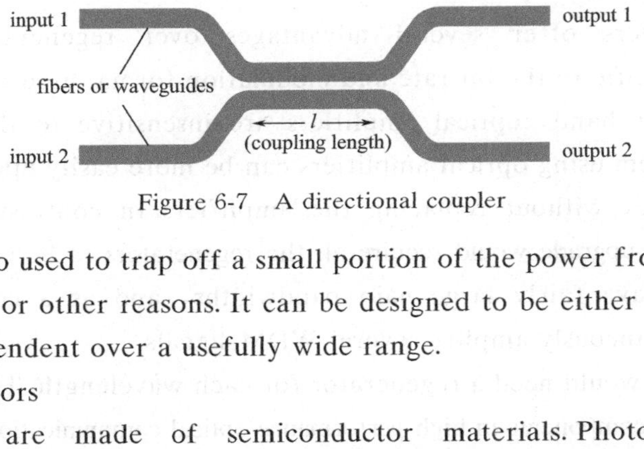

Figure 6-7 A directional coupler

Couplers are also used to trap off a small portion of the power from a light stream for monitoring purposes or other reasons. It can be designed to be either wavelength selective or wavelength independent over a usefully wide range.

(2) Photodetectors

Photodetectors are made of semiconductor materials. Photons incident on a semiconductor are absorbed by electrons in the valence band. As a result, these electrons acquire higher energy and are excited into the conduction band, leaving behind a hole in the valence band. When an external voltage is applied to the semiconductor, these electron-hole pairs give rise to a photocurrent. Photodetectors generate an electrical current proportional to the incident optical power.

The fraction of the energy of the optical signal that is absorbed and gives rise to a photocurrent is called the efficiency η of the photodetector. For transmission at high bit rates over long distances, optical energy is scarce, and thus it is important to design the photodetector to achieve an efficiency η of as close to 1 as possible. This can be achieved by using a semiconductor slab of sufficient thickness. The power absorbed by a semiconductor slab of thickness L can be written as

$$P_{abs} = (1 - e^{-\alpha L})\ P_{in} \tag{6.35}$$

where P_{in} is the incident optical signal power, and α is the absorption coefficient of the material.

Then, we can readily obtain the efficiency of the photodetector

$$\eta = \frac{P_{abs}}{P_{in}} = 1 - e^{-\alpha L} \tag{6.36}$$

The absorption coefficient depends on the wavelength. However, photodetectors have a very wide operating bandwidth since a photodector at some wavelength can also serve as a photodector at all smaller wavelength.

(3) Optical amplifiers

In an optical communication system, the optical signals from the transmitter are attenuated by the optical fiber as they propagate through it. Other optical components such as multiplexers and couplers, also add loss. After some distance, the cumulative loss of signal strength causes the signal to become too weak to be detected. Before this happens, the signal strength has to be restored. Prior to the advent of optical amplifiers over the last decade, the only option was to regenerate the signal, that is, receive the signal and retransmit it. This process is accomplished by regenerators. A regenerator converts the optical signal to an electrical signal, cleans it up, and converts it back into an optical signal for onward transmission.

Optical amplifiers offer several advantages over regenerators. On one hand, regenerators are specific to the bit rate and modulation format used by the communication system. On the other hand, optical amplifiers are insensitive to the bit rate or signal formats. Thus a system using optical amplifiers can be more easily upgraded, for example, to a higher bit rate, without replacing the amplifiers. In contrast, in a system using regenerators, such an upgrade would require all the regenerators to be replaced. Furthermore, optical amplifiers have fairly large gain bandwidths, and as a consequence, a single amplifier can simultaneously amplify several WDM signals.

In contrast, we would need a regenerator for each wavelength. Thus optical amplifiers have become essential components in high-performance optical communication systems. Amplifiers, however, are not perfect devices. They introduce additional noise, and this noise accumulates as the signal passes through multiple amplifiers along its path due to the analog nature of the amplifier. The spectral shape of the gain, the output power, and the transient behavior of the amplifier are also important considerations for system applications. Ideally, we would like to have a sufficiently high output power to meet the needs of the network application. We would also like the gain to be flat over the operating wavelength range and to be insensitive to variations in input power of the signal.

(4) Switches

Optical switches are used in optical networks for a variety of applications. One application of optical switches is in the provisioning of lightpaths, used inside wavelength cross-connects to reconfigure them to support new lightpaths. Another important application is that of protection switching. Here the switches are used to switch the traffic stream from a primary fiber onto another fiber in case the primary fiber fails. The entire operation must typically be completed in several tens of milliseconds, which includes the time to detect the failure, communicate the failure to the appropriate network elements handling the switching, and the actual switch time. Thus the switching time required is on the order of a few milliseconds. In these networks, switches are used to switch signals on a packet-by-packet basis.

Switches are also important components in high-speed optical packet-switched networks. In these networks, switches are used to switch signals on a packet-by-packet basis. For this application, the switching time must be much smaller than the packet duration, and large switches will be needed. For example, ordinary Ethernet packets have lengths between about 60 to 1500 bytes. At 10 Gb/s, the transmission time of a 60 byte packet is 48 ns. Thus, the switching time required for efficient operation is on the order of a few nanoseconds.

Yet another use for switches is as external modulators to turn on and off the data in front of a laser source. In this case, the switching time must be a small fraction of the bit duration. So an external modulator for a 10 Gb/s signal (with a bit duration of 100 ps) must have a switching time (or, equivalently, a rise and fall time) of about 10 ps.

6. 5. 2 Modulations and Demodulations

Our goal in this section is to introduce modulations and demodulations of digital signals.

(1) Modulations

Modulation is the process of converting data in electronic form to optical form for transmission on the fiber.

It can be divided into two categories: direct modulation and external modulation. Direct modulation of the laser or light emitting diode (LED) source can be used for transmission at low bit rates over short distances, whereas external modulation is needed for transmission at high bit rates over long distances.

The most commonly used modulation scheme in optical communication is on-off keying (OOK), which is illustrated in Figure 6-8.

In this modulation scheme, a 1 bit is encoded by the presence of a light pulse in the bit interval or by turning a light source (laser or LED) "on". A 0 bit is encoded (ideally) by the absence of a light pulse in the bit interval or by turning a light source "off". The bit interval is the interval of time available for the transmission of a single bit.

The optical signal emitted by a laser operating in the 1310 or 1550 nm wavelength band

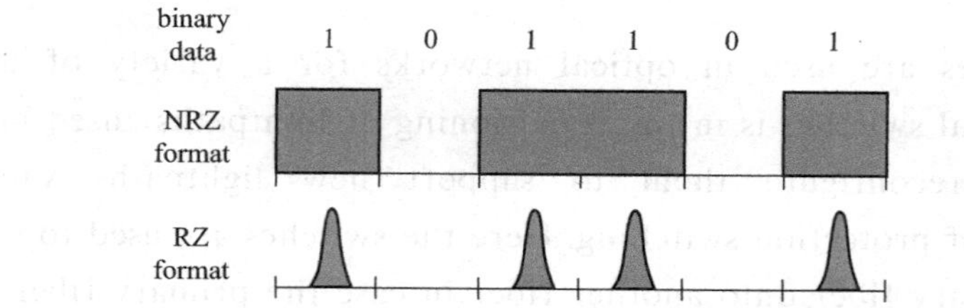

Figure 6-8 On-off keying modulation of binary digital data

has a center frequency around 10^{14} Hz. This frequency is the optical carrier frequency. In what we have studied so far, the data modulates this optical carrier. In other words, with an OOK signal, the optical carrier is simply turned on or off, depending on the bit to be transmitted. Instead of modulating the optical carrier directly, we can have the data first modulate an electrical carrier in the microwave frequency range, typically ranging from 10 MHz to 10 GHz, as shown in Figure 6-9. The upper limit on the carrier frequency is determined by the modulation bandwidth available from the transmitter. The modulated microwave carrier then modulates the optical transmitter. If the transmitter is directly modulated, then changes in the microwave carrier amplitude get reflected as changes in the transmitted optical power envelope, as shown in Figure 6-9.

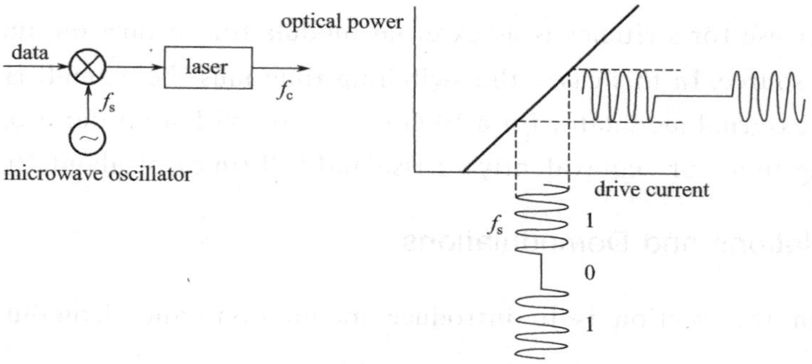

Figure 6-9 Subcarrier modulation

The data stream first modulates a microwave carrier, which, in turn, modulates the optical carrier.

The microwave carrier can itself be modulated in many different ways, including amplitude, phase, and frequency modulation, and both digital and analog modulation techniques can be employed. The figure shows an example where the microwave carrier is amplitude modulated by a binary digital data signal. The microwave carrier is called the subcarrier, with the optical carrier being considered the main carrier. This form of modulation is called subcarrier modulation.

(2) Demodulations

Demodulation is inverse process of modulation, i. e., converting optical signals back into electronic form and extracting the data that is transmitted. The modulated signals are transmitted over the optical fiber where they undergo attenuation and dispersion, have noise added to them from optical amplifiers, and sustain a variety of other impairments.

In principle, the demodulation process can be quite simple. Ideally, it can be viewed as "photon counting". In practice, there are various impairments that are not accounted for by this model. The receiver looks for the presence or absence of light during a bit interval. If no light is seen, it infers that a 0 bit was transmitted, and if any light is seen, it infers that a 1 bit was transmitted. This is called direct detection. Unfortunately, even in the absence of other forms of noise, this will not lead to an ideal error-free system because of the random nature of photon arrivals at the receiver. In practice, most receivers are not ideal, and their performance is not as good as that of the ideal receiver because they must contend with various other forms of noise, as we shall soon see.

The main complication in recovering the transmitted bit is that in addition to the photocurrent due to the signal there are usually three other additional noise currents. The first is the thermal noise current due to the random motion of electrons that is always present at any finite temperature. The second is the shot noise current due to the random distribution of the electrons generated by the photodetection process even when the input light intensity is constant. The shot noise current, unlike the thermal noise current, is not added to the generated photocurrent but is merely a convenient representation of the variability in the generated photocurrent as a separate component. The third source of noise is the spontaneous emission due to optical amplifiers that may be used between the source and the photodetector.

(3) Noise

The most common noises are the front-end amplifier noise and the avalanche photodiodes noise.

We saw that the photodetector is followed by a front-end amplifier. Components within the front-end amplifier, such as the transistor, also contribute to the thermal noise. This noise contribution is usually stated by giving the noise figure of the front-end amplifier. The noise figure F_n is the ratio of the input signal-to-noise ratio (SNR_i) to the output signal-to-noise ratio (SNR_o). Equivalently, the noise figure F_n of a front-end amplifier specifies the factor by which the thermal noise present at the input of the amplifier is enhanced at its output. Thus the thermal noise contribution of the receiver has variance thermal

$$\sigma^2_{\text{thermal}} = \frac{4K_B T}{R_L} F_n B_e \tag{6.37}$$

when the front-end amplifier noise contribution is included. Typical values of F_n are 3~5 dB.

The avalanche gain process in avalanche photodiodes has the effect of increasing the noise current at its output. This increased noise contribution arises from the random nature of the avalanche multiplicative gain, $G_m(t)$. This noise contribution is modeled as an increase in the shot noise component at the output of the photodetector. If we denote the responsivity of the APD by APD, and the average avalanche multiplication gain by G_m, the average photocurrent is given by $\overline{I} = R_{APD} P = G_m RP$, and the shot noise current at the APD output has variance

$$\sigma_{\text{shot}}^2 = 2eG_m^2 F_A(G_m) RPB_e \tag{6.38}$$

The quantity $F_A(G_m)$ is called the excess noise factor of the APD and is an increasing function of the gain G_m. It is given by

$$F_A(G_m) = k_A G_m + (1 - k_A)(2 - 1/G_m) \tag{6.39}$$

The quantity k_A is called the ionization coefficient ratio and is a property of the semiconductor material used to make up the APD. It takes values in the range 0～1. The excess noise factor is an increasing function of k_A, and thus it is desirable to keep k_A small. The value of k_A for silicon (which is used at 0.8 μm wavelength) is 1, and for InGaAs (which is used at 1.3 and 1.55 μm wavelength bands) it is 0.7. Note that $F_A(1) = 1$, and thus Eq. (6.38) also yields the shot noise variance for a pin receiver if we set $G_m = 1$.

(4) Capacity limits of optical fiber

An upper limit on the spectral efficiency and the channel capacity is given by Shannon's theorem. Shannon's theorem says that the channel capacity C for a binary linear channel with additive noise is given by

$$C = B\log_2(1 + S/N) \tag{6.40}$$

Here B is the available bandwidth and S/N is the signal-to-noise ratio. A typical value of S/N is 100. Using this number yields a channel capacity of 350 Tb/s or an equivalent spectral efficiency of 7 b/(s · Hz). Clearly, such efficiencies can only be achieved through the use of multilevel modulation schemes.

In practice, today's long-haul systems operate at high power levels to overcome fiber losses and noise introduced by optical amplifiers. At these power levels, nonlinear effects come into play. These nonlinear effects can be thought of as adding additional noise, which increases as the transmitted power is increased. Therefore they in turn impose additional limits on channel capacity. Recent work to quantify the spectral efficiency, taking into account mostly cross-phase modulation, shows that the achievable efficiencies are of the order of 3～5 b/(s · Hz). Other nonlinearities such as four-wave mixing and Raman scattering may place further limitations. At the same time, we are seeing techniques to reduce the effects of these nonlinearities. Another way to increase the channel capacity is by reducing the noise level in the system. The noise figure in today's amplifiers is limited primarily by random spontaneous emission, and these are already close to theoretically achievable limits. Advances in quantum mechanics may ultimately succeed in reducing these noise limits.

6.5.3 WDM Network Design

There is a clear benefit to build wavelength-routing networks, as opposed to simple point-to-point WDM links. The main benefit is that traffic that is not to be terminated within a node can be passed through by the node, resulting the significant saving in high-layer terminating equipment.

The WDM network provides circuit-switched lightpaths that can have varying degrees of transparency associated with them. Wavelengths can be reused in the network to support multiple lightpaths as long as no two lightpaths are assigned the same wavelength on a given link. Lightpaths may be protected by the network in the event of failures. They can be used to provide flexible interconnections between users of the optical network, such as IP routers, allowing the router topology to be tailored to the needs of the router network.

An optical line terminal (OLT) multiplexes and demultiplexes wavelengths and is used for point-to-point applications. It typically includes transponders, multiplexers, and optical amplifiers. Transponders provide the adaptation of user signals into the optical layer. They also constitute a significant portion of the cost and footprint in an OLT. In some cases, transponders can be eliminated by deploying interfaces that provide already-adapted signals at the appropriate wavelengths in other equipment.

The design of optical network is more complicated than the design of traditional networks. It includes the design of the higher-layer topology (IP or SONET), which is the lightpath topology design problem, and its realization in the optical layer, which is the routing and wavelength assignment problem. These problems may need to be solved in conjunction if the carrier provides IP or SONET VTs over its own optical infrastructure. However, this is difficult to do, and a practical approach may be iteratively solve these problems.

We must concern attention to the wavelength dimensioning problem. The problem here is to provide sufficient capacity on the links of the wavelength-routing network to handle the expected demand for lightpaths. This problem is solved today by periodically forecasting a traffic matrix and (re)designing the network to support the forecasted matrix. Alternatively, you can employ statistical traffic demand models to estimate the required capacities, and we discussed two such models.

New Words and Expressions

access 使用权
accumulate 累积
acoustic 听觉的，声音的，原声的
advent 出现
allocation 分配
analogous 相似的
attenuation 衰减
cladding layer 包层
coefficient 系数
packet 波包
paradigm 示例
pattern 模式，图案
perturbation 烦乱，扰乱
pervasive 普遍的，到处的
resonance 共振
reverse 反面的
spectral 光谱的
topology 拓扑学
versatile 通用的，用途广泛的
vibration 振动
walk-off 离散，走离

Translation

1. Two distinct routes are being followed to overcome this problem.

翻译：为了克服这一问题，正在采取两种不同的方法。

2. Special multisection lasers have been developed to meet the demands of fourth-generation lightwave systems.

翻译：为了满足第四代光波系统的要求，研制了特殊的多截面激光器。

3. Advances in the semiconductor-laser technology play an important role in realizing such high-performance lightwave systems.

翻译：半导体激光器技术的进步对实现这种高性能的光波系统起着重要的作用。

4. In its simplest form, an optical fiber consists of a central core surrounded by a cladding layer whose refractive index is slightly lower than that of the core.

翻译：形式最简单的光纤是由一个中心纤芯和周围的包层组成，包层的折射率略低于纤芯的折射率。

5. Such fibers are generally referred to as step-index fibers to distinguish them from graded-index fibers in which the refractive index of the core decreases gradually from center to the core boundary.

翻译：这种光纤通常被称为阶跃折射率光纤，以区别于渐变折射率光纤，在渐变折射率光纤中，纤芯折射率从中心到纤芯边界逐渐减小。

6. On a fundamental level, the origin of nonlinear response is related to a harmonic motion of bound electrons under the influence of an applied field.

翻译：从根本上讲，非线性响应的起源与在施加电场的影响下束缚电子的谐波运动有关。

7. However, unless special efforts are made to achieve phase matching, the nonlinear processes which involve the generation of new frequencies are not efficient in optical fibers.

翻译：除非做出特殊的努力来实现相位匹配，否则涉及产生新频率的非线性过程在光纤中是无效的。

8. Maxwell's equations can be used to obtain the wave equation that describes light propagation in optical fibers.

翻译：麦克斯韦方程可用于获得描述光在光纤中传播的波动方程。

9. Although the inclusion of radiation modes is crucial in problems involving transfer of power between bounded and radiation modes, they do not play an important role in the discussion of nonlinear effects.

翻译：虽然辐射模式包含在有界模式和辐射模式之间的功率转移问题至关重要，但它们在非线性效应的讨论中并不起重要作用。

10. This is not really the case, unless polarization-maintaining fibers are used, but the approximation works quite well in practice.

翻译：事实并非如此，如果持续使用偏振光纤，该近似值在实际应用中效果良好。

11. Even though operation at the zero-dispersion wavelength is most desirable from the standpoint of pulse broadening, other considerations may preclude such a design.

翻译：即使从脉冲展宽的角度来看，在零色散波长下操作是最理想的，但综合其他考虑可能会排除这种设计。

12. The technique of dispersion management provides a solution to this dilemma.

翻译：分散管理技术提供了解决这一难题的方法。

Unit 7
Nonlinear Optics

7.1 The Definition of Nonlinear Optics

Nonlinear optics is the study of phenomena that occur as a consequence of the modification of the optical properties of a material system by the presence of light. Typically, only laser light is sufficiently intense to modify the optical properties of a material system. The nonlinear response can result in intensity-dependent variation of the propagation characteristics of the radiation fields or in the creation of radiation fields that propagate at new frequencies or in new directions. Nonlinear effects can take place in solids, liquids, gases, and plasmas, and may involve one or more electromagnetic fields as well as internal excitations of the medium. Most of the work done in the field has made use of the high powers available from lasers. The wavelength range of interest generally extends from the far-infrared to the vacuum ultraviolet, but some nonlinear interactions have been observed at wavelengths extending from the microwave to the X-ray ranges.

Nonlinear optics mainly deals with various new optics effects and novel phenomenon arising from the interactions of intense coherent optical radiation with matter.

Before 1960s, in the area of conventional optics many basic mathematical equations or formulae manifested a linear feature.

To show this linear feature of conventional optics, we can consider the following two examples.

First, in order to interpret the refraction, dispersion, scattering, as well as birefringence of light propagation in a medium, we should consider an important physical quantity, the electric polarization induced in the medium. In the regime of conventional optics, the electric polarization vector P is simply assumed to be linearly proportional to the electric field strength E of an applied optical wave, i. e.

$$P = \varepsilon_0 x E \tag{7.1}$$

where ε_0 is the free-space permittivity, x is the susceptibility of a given medium. Based on this linear assumption, Maxwell's equations lead to a set of linear differential equations in which only the terms proportional to the first power of the field E are involved. As a result, there is no coupling between different light beams or between different monochromatic components when they pass through a medium. In other words, if there are several monochromatic optical waves with different frequencies passing through a medium simultaneously, no coherent radiation at any new frequency will be generated.

Second, in conventional optics, the attenuation of an optical beam propagating in an absorptive medium can be described as

$$\frac{dI}{dz} = -\alpha I \tag{7.2}$$

where I is the beam intensity, z is the variable along the propagation direction, and α is a constant for a given medium. The physical meaning of Eq. (7.2) is that the decrease of the beam intensity in a unit propagation length is linearly proportional to the local intensity itself. From Eq. (7.2) we obtain a well-known exponential attenuation expression

$$I(z) = I(0) \cdot e^{-\alpha z} \tag{7.3}$$

This expression implies that for a given propagation length of $z = l$, the transmitted intensity $I(l)$ is linearly proportional to the initial intensity of $I = I(0)$.

So far we have given two examples that manifest a simple linear feature as shown by Eq. (7.1) and Eq. (7.2), respectively. These simple linear assumptions or conclusions given by the conventional optics were widely accepted, and verified by most experimental observation and measurements based in the use of ordinary light sources. However, these situations have been changed radically since the beginning of 1960s.

Shortly after the demonstration of the first laser device (a pulsed ruby laser) in 1960, it was found that these simple linear assumptions or conclusions described above were no longer adequate for circumstances in which an intense laser beam was incident on certain types of optical media. For the sake of clarity, we shall stay with our two examples and show why some higher-order approximations should be employed when an intense laser field interacts with an optical medium.

The first breakthrough was achieved in 1961 when a pulsed laser beam was sent into a piezoelectric crystal sample. In this case researchers, for the first time in the history of optics, observed the second-harmonic generation (SHG) at an optical frequency. Shortly after this discovery, several other coherent optical frequency mixing effects (such as optical sum-frequency generation, optical difference-frequency generation, and optical third harmonic generation) were observed. The researchers realized that all these new effects could be reasonably explained if replaced the linear term on the right-hand side of Eq. (7.1) by a power series

$$P = \varepsilon_0 [x^{(1)} E + x^{(2)} EE + x^{(3)} EEE + \cdots] \tag{7.4}$$

Here, $x^{(1)}$, $x^{(2)}$, and $x^{(3)}$ are the first-order (linear), second-order (nonlinear), and third-order (nonlinear) susceptibility and so on. They are material coefficients and in general are tensors. Substituting Eq. (7.4) into Maxwell's equations leads to a set of nonlinear differential equations that involve high-order-power terms of optical electric field strength; these terms are responsible for various observed coherent optical frequency-mixing effects.

In the same time period, researchers also found that the depletion behavior of an intense laser beam propagating in an absorptive optical medium did not follow the description indicated by Eq. (7.2) or Eq. (7.3). For instance, in a one-photon absorptive medium, if the intensity of the incident beam is high enough, the attenuation coefficient α is no longer a constant and may become a variable that depends on the incident intensity.

Therefore, the exponential attenuation formula like Eq. (7.3) can not be applied and the linear relationship between $I(z = l)$ and $I(0)$ does not hold. In this case, either a saturable absorption or a reverse-saturable absorption effect may take place. Moreover, if there is a two-photon absorption process involved in the medium, the attenuation of an intense incident beam should be described as

$$\frac{dI}{dz} = -\alpha I - \beta I^2 \tag{7.5}$$

where β is the two-photon absorption coefficient, which could be viewed as a constant only if the saturation or reverse-saturation effect can be neglected. In more general cases, if we further extend our consideration to include multiphoton (three-photon or more) absorption processes, then Eq. (7.5) should be generalized to the following form

$$\frac{dI}{dz} = -\alpha I - \beta I^2 - \gamma I^3 - \cdots \tag{7.6}$$

Here γ is the three-photon absorption coefficient and so on.

Based on these comparisons described above, we can conclude that the main concern in conventional optics is the propagation and interaction with matter of the light from ordinary light sources, wherein the intensities of the light beams are so low that even a simple linear approximation is enough to give a good theoretical explanation for the related optical effects and phenomena. In this sense, the conventional optics may also be called "linear optics" or "optics of weak light". On the other hand, "nonlinear optics" mainly deals with the interaction of intense laser radiation with matter. In the latter case, the intensities of laser beams can be so high that a great number of new effects and novel phenomena can be observed, and some high-order nonlinear approximations have to be employed to explain these new effects and phenomena. In this sense, nonlinear optics may also be called "optics of intense light". In general, the contents of nonlinear optics are much more extensive than that of linear optics and, accordingly, the theories of the former are more complicated than that of the latter.

7.2 The History of Nonlinear Optics

The formation of nonlinear optics originated in the early 1960s. The discovery of the optical second-harmonic generation by P. A. Franken in 1961 was commonly recognized as the first milestone of the formation of nonlinear optics. Shortly after that, several other optical frequency-mixing effects were sequentially demonstrated based on the use of laser radiation, which include the optical sum-frequency generation, optical third-harmonic generation, optical rectification, optical difference-frequency generation, and the optical parametric amplification and oscillation. These experimental demonstrations not only verified the validity of nonlinear polarization theories but also provided an alternative approach to generate coherent optical radiation. During the same time period, another important event was the discovery of stimulated Raman scattering (SRS), which can be recognized as the second milestone in the history of nonlinear optics. Later, researchers reported the observation of stimulated Brillouin scattering (SBS), which arose from the interaction of an intense monochromatic optical field with the induced hypersonic field in a scattering medium through the so-called optical electrostriction mechanism. Since then, the stimulated Brillouin scattering has become an efficient technique to generate or amplify the coherent optical radiation with a small frequency-shift or fine tenability.

Another major subject of nonlinear optics is related to the refractive-index change induced by an intense laser beam as well as the impact of this change on the laser beam itself. An important article focused on this issue was published in 1964 with a conceptual discussion and a semi-quantitative description of the self-focusing (self-trapping) behavior of an intense optical beam propagating in a nonlinear medium. Further studies of dynamic self-focusing processes for short laser pulses revealed special properties of the moving-focus as well as the new phenomena, self-phase modulation and spectral self-broadening. Now it is well known that the self-phase modulation and spectral self-broadening effects are among the basic mechanisms for generating ultra-short laser pulses and the continuum radiation with a super-broad spectral band.

Other kinds of nonlinear optical phenomena were also reported in the 1960s, the so-called transient coherent optical effects, include photon echoes, self-induced transparency, and optical nutation. These effects are related to the transient-response behavior of a resonant optical medium interacting with short optical pulses or a fast-switched optical field. Some of them are the optical analog of the corresponding effects in nuclear magnetic resonance studies. The studies of transient optical effect can provide a new approach to investigate the relaxation behavior of resonant transitions in absorptive media.

In addition, there are other two fundamental nonlinear optical effects. The optical saturable absorption effect in organic dye solutions and other materials were well studied and soon applied to the Q-switching and mode-locking of laser devices. The other effect is two-photon absorption (TPA). The earliest experimental demonstrations of TPA-induced fluorescence were achieved with the use of laser radiation. The initial research work of laser-based TPA and the related processes stimulated an extensive investigation on two-photon and multi-photon induced absorption, fluorescence, ionization and dissociation. Later, all these kinds of studies have formed the important part of nonlinear optics and laser spectroscopy.

From 1970 to 1990, accompanying with the great successes in laser science and technology, many new effects were exploited, including the optical bistability (OBIS) soliton, squeezed state, etc.. In the domain of laser technique, dramatic progresses in shortening the pulse duration further to femtosecond (fs) scale have greatly stimulated the researches of ultrafast processes in photophysics, photochemistry and photobiology. The applications of advanced laser techniques have led to the great successes in optical fiber communications, laser manufacturing of microstructures, etc..

Four-wave mixing (FWM) was found in the first decade of the laser epoch. The unique features of FWM are the recovery of the phase and the correction of the phase aberration, which are very attractive in many applications. A special technique is degenerate four-wave mixing (DFWM), where two counter-propagating pump beams are used to generate a new beam in the opposite direction with the probe beam. This new beam has the complex conjugate relation with the probe beam. In such case, the phase aberration introduced by the propagation in atmosphere or other media could be effectively corrected, making DFWM to be very promising in the applications for the laser propagation in long-range distance, and the self-tracking system used for the laser fusion. A special configuration of FWM called coherent anti-Stokes Raman scattering (CARS) was studied during this period. The unique advantage of CARS technique is that it could avoid the interference of the fluorescence in the measurement and improves greatly the signal to noise ratio.

The stimulated electronic Raman scattering (SERS) is the stimulated process happening among the electronic states. SERS was widely studied in various metal vapors, such as Na, K, Rb and Cs. As the energy intervals between the excited states of these atoms are small, the generated SERS emissions are in the near infrared to middle infrared ranges, which are very useful in the research of large molecules and narrow gap semiconductors. In addition to SERS, the stimulated hyper Raman scattering was also observed. But the interest in this kind of research decreased gradually due to the fact that it was hard to meet the demand of the applications with high conversion efficiency.

One of the most important discoveries in this period was the OBIS. In 1957, S. L. McCall et al. at the Bell laboratories observed optical bistable property in the sodium vapor inside a Fabry-Perot cavity. This character is caused by the combination of the resonant cavity and the nonlinear optical property of the medium. After the first observation, a great number of laboratories and universities carried out the research work and observed OBIS in a variety of materials. Different kinds of OBIS were soon discovered, such as absorptive OBIS, dispersive and hybrid OBIS. The great interest on this phenomenon was that it stimulated the human's imagination on manufacturing the optical computer. At the same time, the chaos phenomenon was also carefully studied.

In accordance with the development of the fiber communications, the nonlinear optical research of fibers demonstrated the observation of Stokes SRS, multi-order Stokes and anti-Stokes SRS in fibers. The research on the SRS process in fiber led to the invention of the fiber Raman laser. Even in this isotropic material, SHG in long fiber was generated. Self-phase modulation (SPM) was found to broaden the spectral band of the laser beam propagating inside the fiber. It was realized later that SPM could be used to compensate a new kind of light called soliton.

The first soliton was observed by L. F. Mollenauer in 1980, seven years after the concept of soliton proposed by A. Hasegawa. Other scientists developed the theory of soliton and suggested the construction of soliton laser, which was built soon later. The success in fabricating the soliton laser has attracted much attention for its potential use in the optical communications, as it would decrease the noise and hopefully increase the volume of the date bit-rate.

The progress in quantum optics during these years was very encouraging. In 1985, R. E. Slusher at Bell laboratories reported the observation of a new state, called squeezed state, in sodium vapor by using non-degenerate four-wave mixing (NDFWM)、The special feature of the squeezed state is that it can transfer the noise from one quadratic component to another quadratic component of the optical electric field, thus greatly reduces the noise in the squeezed component. This discovery was found potential in the research of the universe attraction force and in the low-noise optical communications.

For many nonlinear optical researches and applications, especially in those using SHG, SFG and optical parametric oscillation (OPO) effects, the nonlinear optical crystals with large second-order susceptibility are definitely required. For searching new efficient nonlinear optical crystals, it is important to calculate the second-order susceptibility of molecules and crystals. To solve this problem, C. T. Chen and his coworkers at the Chinese Academy of Sciences developed an anion radical theory, which was first successful applied to the calculation of the second-order susceptibilities of $NaNO_2$ and $LiNbO_3$. In 80s of the last century, they got the achievements in the elaborate design and the successful fabrication of new nonlinear optical crystals BBO and LBO, which improve greatly the quality of many nonlinear optical devices and laser systems.

Since 1990, the research in nonlinear optical field has been continuously developing both in theoretical and applicable aspects.

One of the most important progresses is the rapid development of the solid-state ultrashort laser systems, which provide the powerful tools that can be used to explore the ultrafast processes happening in the nature and to study the transient response of the novel materials. The crystals, such as Ti: sapphire, were found to have very broad fluorescence band and self mode-locking features, which make them to be the appropriate gain materials for constructing the ultrashort laser system. Based on the maturity of the solid-state fs laser techniques, the optical parametric amplification has been successfully used to generate the tunable fs laser pulses covering the near infrared, visible and even ultraviolet ranges. The development of fs laser systems also stimulates the research on the generation of the laser pulses with even shorter duration or shorter wavelength, including attosecond pulses and soft X-ray pulses.

The research on the soliton has got big successes in recent years. Both the temporal and spatial solitons have been studied theoretically and experimentally. The various types of solitons, including bright soliton, dark soliton and grey soliton, have been generated and their interactions were studied. The temporal-spatial soliton was observed in the nonlinear optical crystals like $LiIO_3$. As the soliton can keep its temporal profile during the propagation in long distance, which is very important for the optical communication with huge data bit-rate, the soliton communication has been widely studied and the prototype communication system based on the soliton was constructed.

In accordance with the development of the optical communications, different optical devices with ultrafast response have been designed and fabricated, including the optical switching, optical modulator and detectors. The design of the optical switching based on the photonic crystals and the combination of the cascade second-order processes with semiconductor quantum wells have been proposed and studied experimentally. All these researches will certainly improve the qualities of the optical component and increase the data bit-rate for the optical communication.

During this period, there are still many research work and progresses on the pulse compression, the quantum optics, the time-resolved spectroscopic, etc..

The developments in the past nearly sixty years have proved that nonlinear optics is a so exciting and fruitful research brand, which provides big progress in study the materials and systems in various scientific fields. It can be expected that continuous development of nonlinear optics will bring new achievement and successes for science and technology.

7.3 The Features of Interaction of Intense Light with Matter

Regarding the interaction of light radiation with matter, the total number of newly discovered effects and phenomena after the advent of lasers is even larger than that before the invention of lasers. One may ask why so many new things can be found in a so short time period (only five to six decades). To answer this question, we should consider the essential differences between the light beams from laser devices and that from ordinary light sources. Only based on these differences, we can realize how powerful the laser radiation could be when it interacts with matter.

As mentioned in Chapter 5, the laser radiation is generated based in stimulated emission from a population inversion system, whereas the ordinary light is based on spontaneous emission from conventional light sources. Consequently, these two emission mechanisms lead to a great difference in the parameters used to describe the properties of light radiation.

The following are the common parameters to characterize a quasi-directional and quasi-monochromatic light field.

① Intensity is defined as

$$I = \frac{P}{S} \tag{7.7}$$

where P is the total light power (in units of watt) and S is the cross section of the light beam (in units of m^2 or cm^2). The unit of the intensity is W/m^2 or W/cm^2.

② Spectral intensity is defined as

$$I(\nu) = \frac{P}{S\Delta\nu} \tag{7.8}$$

where $\Delta\nu$ is the spectral width of the light radiation (in units of hertz), the unit of spectral intensity is $W/(cm^2 \cdot Hz)$.

③ Brightness is defined as

$$B = \frac{P}{S\Omega} = \frac{1}{\Omega} \tag{7.9}$$

where Ω is the divergent solid angle of the light beam (in unit of steradian). The unit of brightness is W/ (cm^2 · sr).

④ Spectral brightness is defined as

$$B(\nu) = \frac{P}{S\Omega\Delta\nu} = \frac{I(\nu)}{\Omega} \tag{7.10}$$

The unit of the brightness is W/ (cm^2 · sr · Hz).

⑤ Photon degeneracy is defined as the average photon number contained in a single mode of optical field. This parameter is the basic quantity to describe the photon field in quantum electrodynamics and can be determined in the following way. For a quasi-directional and quasi-monochromatic light radiation the total photon number passing through a given beam section of S within a given time interval of Δt is

$$F = \frac{P\Delta t}{h\nu} \tag{7.11}$$

where $h\nu$ is the energy of a single photon, h is Planck constant, ν is the frequency of light. On the other hand, the mode number (or phase-sell number) associated with the above F photons is given by

$$N = \frac{\Delta t}{\delta t}\frac{S}{\delta S} = \frac{\Delta t}{1/\Delta\nu} \cdot \frac{S}{\lambda^2/\Omega} \tag{7.12}$$

Here $\delta t = 1/\Delta\nu$ is the longitudinal coherent time of the optical radiation, $\Delta\nu$ is the spectral width, $\delta S = \lambda^2/\Omega$ is the coherent section of the light beam, Ω is the solid angle of beam divergence. Assuming the light beam is nonpolarized, there should be two independent polarization states; thus the photon degeneracy $\overline{n}$ can be finally determined by

$$\widetilde{n}(\nu) = \frac{F}{2N} = \frac{P}{\left(\frac{2h\nu}{\lambda^2}\right) S\Omega\Delta\nu} \tag{7.13}$$

The photon degeneracy is a dimensionless quantity. From Eq. (7.7) to Eq. (7.10), one can see that the light intensity represents the power density, the spectral intensity represents the power density within a unit spectral interval, the brightness represents the power density within a unit solid angle and the spectral brightness represents the power density within a unit solid angle and a unit spectral interval, respectively. In addition, comparing Eq. (7.10) with Eq. (7.13) one can see that there is only a difference of factor $(2h\nu/\lambda^2)^{-1}$ between the spectral brightness $B(\nu)$ and the photon degeneracy $\overline{n}(\nu)$. Therefore, they can be viewed as two equivalent quantities. According to conventional optics, the brightness of a light beam cannot be increased by passing it through any kinds of optical imaging or transmission systems. It can also be expressed in terms of quantum statistics that the total number of modes for a given photon ensemble cannot be compressed by any ordinary optical systems; therefore, the photon degeneracy cannot be increased by any types of ordinary optical devices. However, these two equivalent conclusions are no longer valid for lasers and nonlinear optical devices. It is well known that the brightness or photon degeneracy of a weak optical signal can be dramatically increased based on the coherent amplification through a lasing medium, a stimulated scattering medium, or an optical parametric amplifier system. For a laser oscillator system, the number of the total lasing modes can be greatly restricted by choosing appropriate cavity configurations and mode selection techniques. As a result, the photo degeneracy of the output laser beam can be extremely high.

According to the electromagnetic theory of light, on the other hand, the spectral intensity of a quasi-parallel laser beam is equal to the magnitude of the Poynting's vector of a monochromatic plane electromagnetic wave, i. e.,

$$I(\nu) = \frac{1}{2}\varepsilon_0 c n_0 |E(\nu)|^2 \tag{7.14}$$

where c is the speed of light in vacuum, n_0 is the linear refractive index of the medium, and $E(\nu)$ is the electric field strength of the monochromatic plane wave. From Eq. (7.8) and Eq. (7.14) one can see that the values of $I(\nu)$ and $E(\nu)$ can be significantly increased when the beam size of the light radiation is compressed by using a reverse beam-expander or a focusing optical system, which are often employed in the experimental studies of nonlinear optics.

In addition, the values of $I(\nu)$ and $E(\nu)$ of a laser beam can be further increased by using an optical focusing system as described above. In the sense of radiation potentials in interacting with matter and creating various nonlinear responses, the differences between ordinary light and laser light are mostly similar to the differences between conventional weapons and strategic nuclear weapons. Therefore, we can say that the light from ordinary light sources is a weak optical radiation characterized by $\overline{n}$ and can only create an extremely low electric field strength. Based on this reason, all the nonlinear terms in the expression of the polarization [see Eq. (7.4)] can be neglected. In contrast, the laser radiation is an intense coherent light characterized by $\overline{n} \gg 1$ and can provide a much stronger optical-frequency electric field, which can even be comparable with the internal electric field of the atom or molecule. In such a case, the nonlinear terms of the polarization expression cannot be entirely neglected and may play a vital role for various nonlinear optical effects. Based on quantum statistics, a light radiation with $\overline{n} \ll 1$ manifests the feature of a shot noise field when it interacts with matter; whereas a radiation with $\overline{n} \gg 1$ manifests the feature of a coherent wave field when it interacts with matter. That is an alternative insight to understand why so many nonlinear optical effects (especially the coherent wave mixing effects in nonlinear media) can be observed only by using laser radiation but not ordinary optical radiation.

In summary, on the one hand, the parameters of laser radiation (such as power, beam divergence, pulse duration, wavelength, spectral width, polarization status, etc.) can be well controlled or modified based on existing laser techniques. On the other hand, there is a great variety of nonlinear optical media, which can be various materials (inorganic, organic, biological, etc.), in different physical states (solid, liquid, gas, plasma, liquid crystal, etc.), and with different reaction centers (molecules, atoms, ions, atomic nuclei, electrons, color centers, photons, excitons, plasmons, etc.). So that, it is not surprising that so many new effects and novel phenomena in nonlinear optics have been found within only four to five decades since 1960s. These effects and phenomena are related to the intense light induced opto-optical, opto-electric, opto-magnetic, opto-acoustic, opto-thermal, opto-mechanical, opto-chemical, and opto-biological interactions in optical media. Generally speaking, all these kinds of interactions can be used to develop various new techniques that may provide many advantages, such as high efficiency, high resonant selectivity, high spectral resolution, high temporal resolution, high spatial resolution, and high sensitivity.

7.4 The Theory Framework of Nonlinear Optics

Basically, two major theoretical approaches can be employed in nonlinear optics as well as in laser physics. The first is the semi-classical theory, and the second is the quantum electrodynamical theory. The most essential feature of the semi-classical theory is that the media composed of atoms or molecules are described by the theory of quantum mechanics, while the light radiation is described by the classical Maxwell's theory. The key issue of semi-classical theory in nonlinear optics is to give the expressions of macroscopic nonlinear electric polarization for optical media. For this purpose, the density matrix method, which is a special approach based on both quantum mechanics and statistical physics, is used to derive the expressions for various orders of electric susceptibilities as $x^{(1)}$, $x^{(2)}$, $x^{(3)}$, ..., and the expressions for various orders of polarization components as $P^{(1)}$, $P^{(2)}$, $P^{(3)}$, ..., and so on. Substituting the appropriate nonlinear polarization components into the generalized wave equations, we are able, in principle, to predict many possible nonlinear optical responses of the medium for a given condition of the input intense optical field(s).

In contrast, the quantum theory of radiation in the regime of quantum electrodynamics treats the medium and optical field as a combined and quantized system. In other words, both the medium and the optical field should be described in the way of quantum mechanics. As a result, the wave function of the combined system is expressed as the product of the eigen function of a molecular system and the eigen function of a quantized photon field. In this case, the key issue is to determine the probability of state change of the combined system due to interaction between the photon field and the medium. Usually, the state changes of the combined system are related to the transition of molecular system from its initial eigen state to the final state and the simultaneous changes of the photon numbers among different photon mode.

It should be pointed that there is no major contradiction or inconsistency between the results and conclusions given by these two theoretical approaches. In fact, they can give the same quantitative results in many cases, such as the cross section of Raman scattering as well as the cross section of two-photon absorption. Nevertheless, these two theoretical approaches have their own usefulness and shortcomings. In this sense, these two different theoretical regimes are parallel and complementary to each other in the scope of nonlinear optics.

The most successful example of the semi-classical theory in nonlinear optics is the derivation of quantitative expressions for various orders of nonlinear susceptibilities of optical media. A semi-classical theoretical approach can be employed to explain all those nonlinear optical effects and phenomena, such as various coherent optical wave-mixing effects. Nevertheless, there are some limitations inherently associated with the semi-classical theory. First, this theory cannot distinguish the difference between the stimulated and the spontaneous processes of radiation, scattering, and parametric photon emission. In order to describe the spontaneous processes, the correspondence principle has to be invoked in the semi-classical regime. For example, Einstein's coefficient relation has to be used to describe the difference between the probabilities of spontaneous emission and stimulated emission. Second, some important physical facts (such as transition relaxation and spectral linewidth) can only be considered by introducing a phenomenological damping factor into the equation of density matrix. Finally, there are a number of nonlinear optical effects (such as stimulated Raman scattering, SBS, CARS process, two-photon absorption, third-harmonic generation, as well as induced refractive index change), all of them can be described with a nominal third-order nonlinear susceptibility $x^{(3)}$. In these cases, however, the nonlinear polarization theory can not reveal the essential difference in origins and mechanisms of those entirely different effects. As a result of that failure, sometimes one may find confusion and terminological ambiguity in classification and description of some nonlinear optical processes within the regime of semi-classical theory.

In quantum electrodynamics, the quantum theory of radiation is a more rigorous theoretical approach that, in principle, can be perfectly used to explain or describe any kinds of effects and phenomena related to the interaction of radiation field with matter in both qualitative and quantitative ways. There are many well-known examples that have shown the advantage of the quantum theory of radiation over the semi-classical theory. First, the relationship between stimulated emission (or scattering) and spontaneous emission (or scattering) can be naturally derived without the need of using the correspondence principle. Second, the selection rule, life-time of state, and spectral linewidth can be quantitatively determined for a given molecular system. Third, the conversation of energy and momentum between the photon field and molecular system can be logically applied to various nonlinear optical processes without the need of using the so-called Manley-Rowe relation (for conservation of energy) and the phase-matching requirement (for conversation of momentum). Finally, the most important feature of the quantum theory of radiation is that a concept of virtual energy level can be introduced, which represents an intermediate quantum state occupied by the combined system of the photon field and the medium. Based on the concept of virtual energy level or intermediate state, the principles and mechanisms of most major nonlinear optical effects can be consistently interpreted and, in many cases, clearly illustrated by an energy-level diagram involving the transitions via virtual energy levels. On the other hand, however, the mathematics derivation in the regime of quantum electrodynamics is rather lengthy and cumbersome specific issues. Therefore, in practice the all-quantum derivations of the related formulae are only applied in those cases where the semi-classical approach is obviously poorly or failed.

7.5 The Descriptions of Nonlinear Optical Processes

In the present section, we present brief qualitative descriptions of a number of nonlinear optical processes. In addition, for those processes that can occur in a lossless medium, we indicate how they can be described in terms of the nonlinear contributions to the polarization described by Eq. (7.4). Our motivation is to provide an indication of the variety of nonlinear optical phenomena that can occur. In this section we also introduce some notational conventions and some of the basic concepts of nonlinear optics.

7.5.1 Second-Harmonic Generation

As an example of a nonlinear optical interaction, let us consider the process of second-harmonic generation, which is illustrated schematically in Figure 7-1.

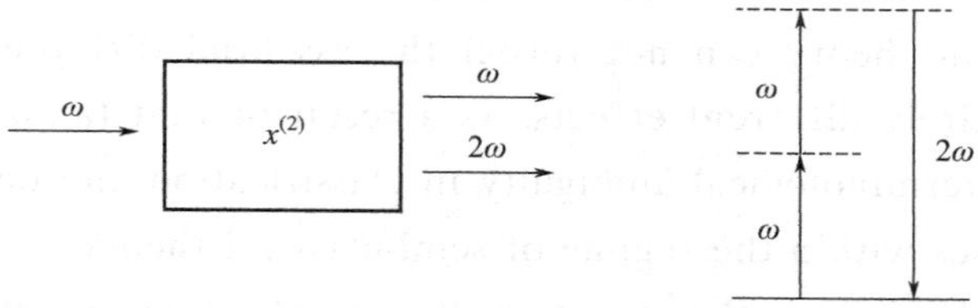

(a) geometry of SHG (b) energy-level diagram describing SHG

Figure 7-1 Second-harmonic generation

Here a laser beam whose electric field strength is represented as

$$\tilde{E}(t) = E\mathrm{e} - i\omega t + \mathrm{c.\,c.} \tag{7.15}$$

is incident upon a crystal for which the second-order susceptibility $x^{(2)}$ is nonzero. The nonlinear polarization that is created in such a crystal is given according to Eq. (7.4) as

$$\tilde{P}^2(t) = \varepsilon_0 x^{(2)} \tilde{E}^2(t)$$

or explicitly as

$$\tilde{P}^2(t) = 2\varepsilon_0 x^{(2)} EE^* + (\varepsilon_0 x^{(2)} \tilde{E}^2 \mathrm{e}^{-i2\omega t} + \mathrm{c.\,c.}) \tag{7.16}$$

We see that the second-order polarization consists of a contribution at zero frequency (the first term) and a contribution at frequency 2ω (the second term).

$$\nabla^2 \tilde{E} - \frac{n^2 \partial^2 \tilde{E}}{C^2 \partial t^2} = \frac{1}{\varepsilon_0 c^2} \frac{\partial^2 \tilde{P}_{\mathrm{NL}}}{\partial t^2} \tag{7.17}$$

According to Eq. (7.17), this latter contribution can lead to the generation of radiation at the second-harmonic frequency. Note that the first contribution in Eq. (7.16) does not lead to the generation of electromagnetic radiation (because its second time derivative vanishes). It leads to a process known as optical rectification, in which a static electric field is created across the nonlinear crystal.

Under proper experimental conditions, the process of second-harmonic generation can be so efficient that nearly all of the power in the incident beam at frequency ω is converted to radiation at the second-harmonic frequency 2ω. One common use of second-harmonic generation is to convert the output of a fixed-frequency laser to a different spectral region. For example, the Nd: YAG laser operates in the near infrared at a wavelength of 1.06 μm. Second-harmonic generation is routinely used to convert the wavelength of the radiation to 0.53 μm, in the middle of the visible spectrum.

Second-harmonic generation can be visualized by considering the interaction in terms of the exchange of photons between the various frequency components of the field. According to this picture, which is illustrated in part (b) of Figure 7-1, two photons of frequency ω are destroyed, and a photon of frequency 2ω is simultaneously created in a single quantum-mechanical process. The solid line in the figure represents the atomic ground state, and the dashed lines represent what are known as virtual levels. These levels are not energy eigen levels of the free atom but rather represent the combined energy of one of the energy eigen states of the atom and of one or more photons of the radiation field.

7.5.2 Frequency Mixing Generation

Let us next consider the circumstance in which the optical field incident upon a second-order nonlinear optical medium consists of two distinct frequency components, which we represent in the form

$$\tilde{E}(t) = E_1 e^{-i\omega_1 t} + E_2 e^{-i\omega_2 t} + \text{c. c.} \tag{7.18}$$

Then, assuming as in Eq. (7.4) that the second-order contribution to the nonlinear polarization is of the form

$$\tilde{P}^2(t) = \varepsilon_0 x^{(2)} \tilde{E}(t)^2 \tag{7.19}$$

We find that the nonlinear polarization is given by

$$\tilde{P}^2(t) = \varepsilon_0 x^{(2)} [E_1^2 e^{-2i\omega_1 t} + E_2^2 e^{-2i\omega_2 t} + 2E_1 E_2 e^{-i(\omega_1 + \omega_2)t} + 2E_1 E_2{}^* e^{-i(\omega_1 - \omega_2)t} + \text{c. c.}] \tag{7.20}$$

It is convenient to express this result using the notation

$$\tilde{P}^2(t) = \sum_n P(\omega_n) e^{-i\omega_n t} \tag{7.21}$$

where the summation extends over positive and negative frequencies ω_n. The complex amplitudes of the various frequency components of the nonlinear polarization are hence given by

$$P(2\omega_1) = \varepsilon_0 x^{(2)} E_1^2 (\text{SHG})$$
$$P(2\omega_2) = \varepsilon_0 x^{(2)} E_2^2 (\text{SHG})$$

$$P(\omega_1+\omega_2)=2\varepsilon_0 x^{(2)} E_1 E_2 \text{(SHG)}$$

$$P(\omega_1-\omega_2)=2\varepsilon_0 x^{(2)} E_1 E_2^* \text{(DFG)}$$

$$P(0)=2\varepsilon_0 x^{(2)} (E_1 E_1^* + E_2 E_2^*) E_1 E_2^* \text{(OR)} \tag{7.22}$$

Here we have labeled each expression by the name of the physical process that it describes, such as second-harmonic generation (SHG), sum-frequency generation (SFG), difference-frequency generation (DFG), and optical rectification (OR). Note that, in accordance with our complex notation, there is also a response at the negative of each of the nonzero frequencies just given

$$P(-2\omega_1)=\varepsilon_0 x^{(2)} E_1^{*2}$$

$$P(-2\omega_2)=\varepsilon_0 x^{(2)} E_2^{*2}$$

$$P(-\omega_1-\omega_2)=2\varepsilon_0 x^{(2)} E_1^* E_2^*$$

$$P(\omega_2-\omega_1)=2\varepsilon_0 x^{(2)} E_2 E_1^* \tag{7.23}$$

However, since each of these quantities is simply the complex conjugate of one of the quantities given in Eq. (7.22), it is not necessary to take explicit account of both the positive and negative frequency components.

We see from Eq. (7.22) that four different nonzero frequency components are present in the nonlinear polarization. However, typically no more than one of these frequency components will be present with any appreciable intensity in the radiation generated by the nonlinear optical interaction. The reason for this behavior is that the nonlinear polarization can efficiently produce an output signal only if a certain phase-matching condition is satisfied, and usually this condition cannot be satisfied for more than one frequency component of the nonlinear polarization. Operationally, one often chooses which frequency component will be radiated by properly selecting the polarization of the input radiation and the orientation of the nonlinear crystal.

7.5.3 Sum-Frequency Generation

Let us now consider the process of sum-frequency generation, which is illustrated in Figure 7-2. According to Eq. (7.22), the complex amplitude of the nonlinear polarization describing this process is given by the expression

$$P(\omega_2+\omega_1)=2\varepsilon_0 x^{(2)} E_2 E_1 \tag{7.24}$$

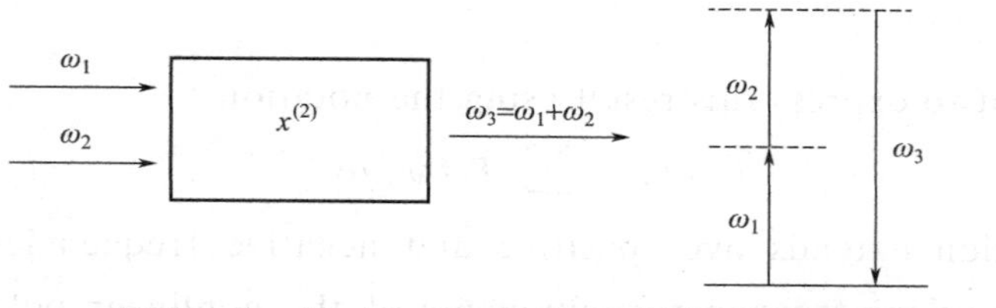

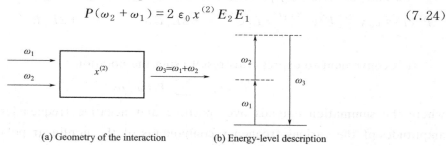

(a) Geometry of the interaction (b) Energy-level description

Figure 7-2 Sum-frequency generation

In many ways the process of sum-frequency generation is analogous to that of second-harmonic generation, except that in sum-frequency generation the two input waves are at different frequencies. One application of sum-frequency generation is to produce tunable radiation in the ultraviolet spectral region by choosing one of the input waves to be the output of a fixed-frequency visible laser and the other to be the output of a frequency-tunable visible laser.

7.5.4 Difference-Frequency Generation

The process of difference-frequency generation is described by a nonlinear polarization of the form

$$P(\omega_1 - \omega_2) = 2\varepsilon_0 x^{(2)} E_1 E_2^* \tag{7.25}$$

and is illustrated in Figure 7-3. Here the frequency of the generated wave is the difference of those of the applied fields. Difference-frequency generation can be used to produce tunable infrared radiation by mixing the output of a frequency-tunable visible laser with that of a fixed-frequency visible laser.

Superficially, difference-frequency generation and sum-frequency generation appear to be very similar processes. However, an important difference between the two processes can be deduced from the description of difference-frequency generation in terms of a photon energy-level diagram [Figure 7-3 (b)]. We see that conservation of energy requires that for every photon that is created at the difference frequency $\omega_3 = \omega_1 - \omega_2$, a photon at the higher input frequency (ω_1) must be destroyed and a photon at the lower input frequency (ω_2) must be created. Thus, the lower frequency input field is amplified by the process of difference-frequency generation. For this reason, the process of difference-frequency generation is also known as optical parametric amplification. According to the photon energy-level description of difference-frequency generation, the atom first absorbs a photon of frequency ω_1 and jumps to the highest virtual level. This level decays by a two-photon emission process that is stimulated by the presence of the ω_2 field, which is already present. Two-photon emission can occur even if the ω_2 field is not applied. The generated fields in such a case are very much weaker, since they are created by spontaneous two-photon emission from a virtual level. This process is known as parametric fluorescence.

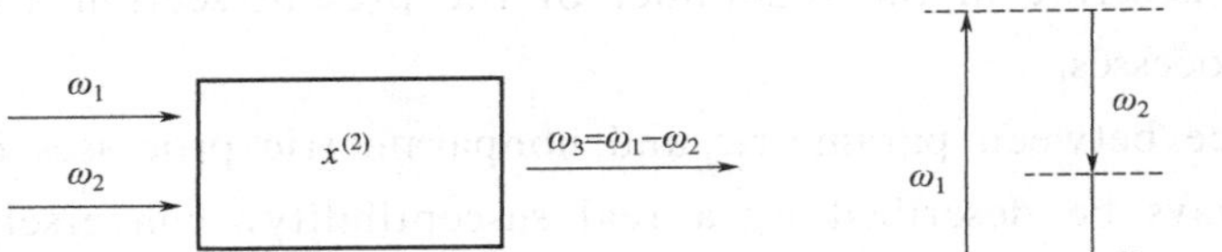

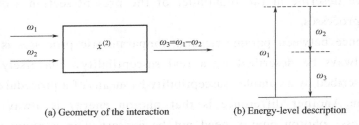

(a) Geometry of the interaction (b) Energy-level description

Figure 7-3 Difference-frequency generation

7.5.5 Optical Parametric Oscillation

We have just seen that in the process of difference-frequency generation the presence of radiation at frequency ω_2 or ω_3 can stimulate the emission of additional photons at these frequencies. If the nonlinear crystal used in this process is placed inside an optical resonator, as shown in Figure 7-4, the ω_2 and/or ω_3 fields can build up to large values.

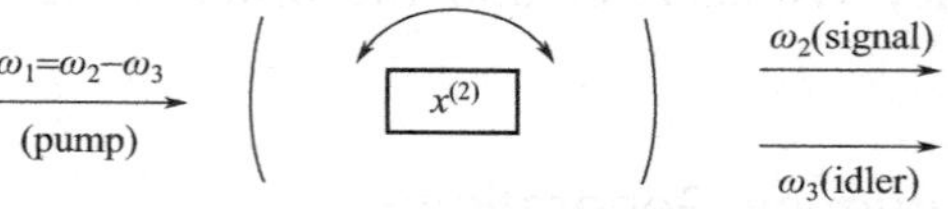

Figure 7-4 The optical parametric oscillator

The cavity end mirrors have high reflectivities at frequencies ω_2 and/or ω_3.

The output frequencies can be tuned by means of the orientation of the crystal.

Such a device is known as an optical parametric oscillator. Optical parametric oscillators are frequently used at infrared wavelengths, where other sources of tunable radiation are not readily available. Such a device is tunable because any frequency ω_2 that is smaller than ω_1 can satisfy the condition $\omega_2+\omega_3=\omega_1$ for some frequency ω_3. In practice, one controls the output frequency of an optical parametric oscillator by adjusting the phase matching condition. The applied field frequency ω_1 is often called the pump frequency, the desired output frequency is called the signal frequency, and the other, unwanted, output frequency is called the idler frequency.

7. 5. 6 Parametric versus Nonparametric Processes

All of the processes described thus far in this chapter are examples of what are known as parametric processes. The origin of this terminology is obscure, but the word parametric has come to denote a process in which the initial and final quantum-mechanical states of the system are identical. Consequently, in a parametric process population can be removed from the ground state only for those brief intervals of time when it resides in a virtual level. According to the uncertainty principle, population can reside in a virtual level for a time interval of the order of $h/\delta E$, where δE is the energy difference between the virtual level and the nearest real level. Conversely, processes that do involve the transfer of population from one real level to another are known as nonparametric processes. The processes that we describe in the remainder of the present section are all examples of nonparametric processes.

One difference between parametric and nonparametric processes is that parametric processes can always be described by a real susceptibility; conversely, nonparametric processes are described by a complex susceptibility by means of a procedure described in the following section. Another difference is that photon energy is always conserved in a parametric process; photon energy need not be conserved in a nonparametric process, because energy can be transferred to or from the material medium. For this reason, photon energy level diagrams of the sort shown in Figure 7-1～Figure 7-3, Figure 7-5, and Figure 7-7 to describe parametric processes play a less definitive role in describing nonparametric processes.

As a simple example of the distinction between parametric and nonparametric processes, we consider the case of the usual (linear) index of refraction. The real part of the refractive index describes a response that occurs as a consequence of parametric processes, whereas the imaginary part occurs as a consequence of nonparametric processes. This conclusion holds because the imaginary part of the refractive index describes the absorption of radiation, which results from the transfer of population from the atomic ground state to an excited state.

7. 5. 7 Saturable Absorption

One example of a nonparametric nonlinear optical process is saturable absorption. Many material systems have the property that their absorption coefficient decreases when measured using high laser intensity. Often the dependence of the measured absorption coefficient α on the intensity I of the incident laser radiation is given by the expression

$$\alpha = \frac{\alpha_0}{1 + I/I_S} \tag{7.26}$$

where α_0 is the low-intensity absorption coefficient, and I_S is a parameter known as the saturation intensity.

One consequence of saturable absorption is optical bistability. One way of constructing a bistable optical device is to place a saturable absorber inside a Fabry-Perot resonator, as illustrated in Figure 7-5. As the input intensity is increased, the field inside the cavity also increases, lowering the absorption that the field experiences and thus increasing the field intensity still further. If the intensity of the incident field is subsequently lowered, the field inside the cavity tends to remain large because the absorption of the material system has already been reduced. A plot of the input-versus-output characteristics thus looks qualitatively like that shown in Figure 7-6. Note that over some range of input intensities more than one output intensity is possible.

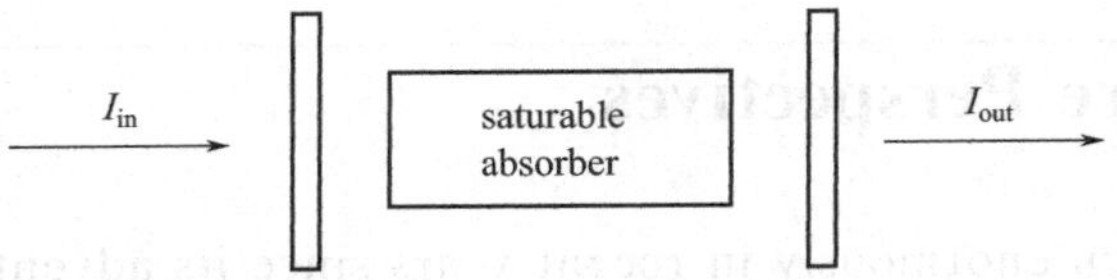

Figure 7-5 Bistable optical device

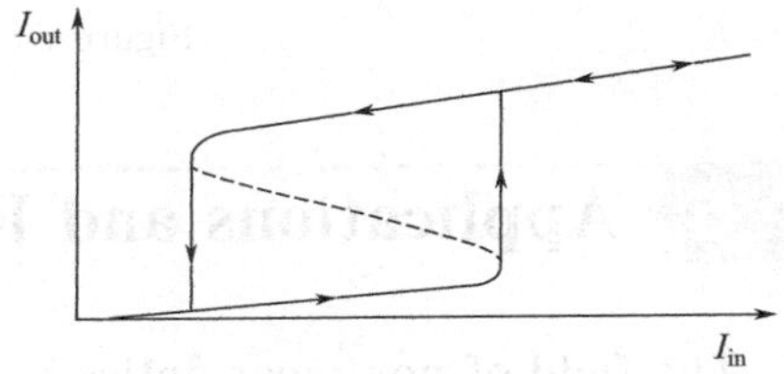

Figure 7-6 Typical input-versus-output characteristics of a bistable optical device

7. 5. 8 Two-Photon Absorption

In the process of two-photon absorption, which is illustrated in Figure 7-7, an atom makes a transition from its ground state to an excited state by the simultaneous absorption of two laser photons. The absorption cross section σ describing this process increases linearly with laser intensity according to the relation

$$\sigma = \sigma^{(2)} I \qquad (7.27)$$

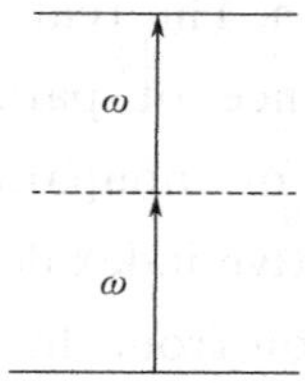

Figure 7-7 Two-photon absorption

where $\sigma^{(2)}$ is a coefficient that describes strength of the two-photon absorption process. (Recall that in conventional, linear optics the absorption cross section σ is a constant.) Consequently, the atomic transition rate R due to two-photon absorption scales as the square of the laser intensity. To justify this conclusion, we note that $R = \sigma I / h\omega$, and consequently that

$$R = \frac{\sigma^{(2)} I^2}{h\omega} \qquad (7.28)$$

Two-photon absorption is a useful spectroscopic tool for determining the positions of energy levels that are not connected to the atomic ground state by a one-photon transition.

7.5.9 Stimulated Raman Scattering

In stimulated Raman scattering, which is illustrated in Figure 7-8, a photon of frequency ω is annihilated and a photon at the Stokes shifted frequency $\omega_B = \omega - \omega_V$ is created, leaving the molecule (or atom) in an excited state with energy $h\omega_V$. The excitation energy is referred to as ω_V because stimulated Raman scattering was first studied in molecular systems, where $h\omega_V$ corresponds to a vibrational energy. The efficiency of this process can be quite large, with often 10% or more of the power of the incident light being converted to the Stokes frequency. In contrast, the efficiency of normal or spontaneous Raman scattering is typically many orders of magnitude smaller.

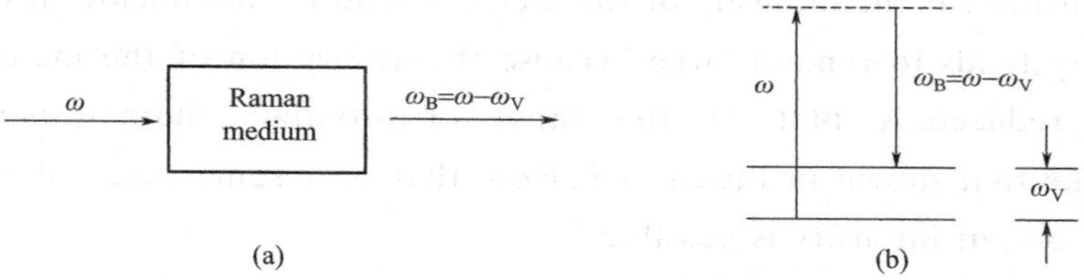

Figure 7-8 Stimulated Raman scattering

7.6 Applications and Future Perspectives

The field of nonlinear optics has grown enormously in recent years since its advent in the early 1960s, soon after the invention of the laser. Nowadays, nonlinear optics has evolved into many different branches, depending on the form of the material used for studying the nonlinear phenomena. The growth of research in nonlinear optics is closely linked to the rapid technological advances that have occurred in related fields, such as ultrafast phenomena, fiber optics, and optical communications. Nonlinear-optics activities range from the fundamental studies related to the interaction between matter and radiation to the development of the devices, components, and systems of tremendous commercial interest for widespread technologies such as optical telecommunications, medicine, and biology.

The application of nonlinear optics that has experienced the most dramatic technological development and economic impact in recent years is definitely related to the design of modern optical communication systems. On the one hand, the nonlinear response of optical fibers leads to impairments in the signal transmission quality in long-haul fiber-optics links. For example, the input power in a fiber is limited by the onset of stimulated Brillouin or Raman scattering. On the other hand, fiber nonlinearity can be exploited to advantage for counteracting the dispersive pulse broadening through the concept of optical solitons, or for compensating fiber losses through the use of stimulated Raman scattering. Currently, hot topics that are included in this issue include the interaction among solitons that belong to different wavelength-multiplexed channels, the coupling of nonlinearity, and polarization-mode dispersion, and the interplay between Raman scattering and solitons. Many designs about parametric and Raman-amplification techniques provide wide-band amplification in addition or as an alternative to the use of erbium-doped fiber amplifiers (EDFAs).

Since the nearly instantaneous response of optical fibers permitting single-channel bit rates as high as 1 Tb/s, the present bottleneck to the channel capacity is set by the limited speed of electronics required for modulation and switching of the information. This issue motivates the research and development of nonlinear optical devices for all-optical data processing and wavelength conversion, such as supercontinuum generation in microstructure fibers, all-optical packet switching and nonlinear coupling in waveguides and resonators, signal processing and spatial soliton propagation in semiconductor microrings and in photorefractive materials. The availability of wavelength conversion functionality will be of great relevance in future all-optical networks. The conversion methods depend on the nonlinear optical material, range from second-harmonic generation (SHG) to quasi-phase matching in optical wave-guides to semiconductor optical amplifiers and highly nonlinear fibers.

The future always belongs to the younger generation. Since the field continues to attract bright students, it may be expected that unexpected new developments will occur. Generally, history is highly nonlinear, and the course of revolutions and the rise and fall of empires is difficult to predict. The future course of nonlinear optics is also unpredictable.

Future technological development should be based on current science. All-optical picosecond switching and quantum logic with entangled states has been demonstrated. Entangled states of two photons are produced, for example, by parametric down conversion of coherent beam or by entangled polarization of two photons emitted in an S-state to S-state transition. Entangled one photon-one atom states have been achieved by Haroche et al. in 1999.

One may envision ultrafast optical supercomputers based on these developments. Two-dimensional (2-D) and three-dimensional (3-D) photonic bandgap materials may be helpful in the manipulation of light beams in all-optical or integrated optical-electronic switching devices.

Three-dimensional holographic information storage has been around for several decades but has yet to achieve large-scale applications. The optical storage of compact disks could be extended to several layers in the perpendicular direction. Fluorescence from centers activated by absorption of two photons from different beams or submicroscopic damage spots from strongly focused femtosecond pulses could, in principle, lead to high-density storage of bits of information.

In 1979, Tajima and Dawson proposed the acceleration of relativistic electrons in the wake field of space charges in the plasmas created by laser pulses with large spatial and temporal gradients. The technological development of powerful femtosecond pulse generators may lead to a new type of electron accelerator. Such devices would be much more compact and presumably cheaper than an extension of the current linear accelerator. High-power femtosecond laser pulses can propagate in the atmosphere over considerable distances. A computational model, which includes self-focusing, multi-photon ionization, the formation of plasma blobs with large spatial and temporal gradients, and the diffraction of light by these plasma blobs, describes the formation of a dynamic, turbulent light guide.

As to future science, our predictions are limited to extensions of currently active areas. Short X-ray pulses have already been obtained by femtosecond excitation of small targets. Perhaps attosecond spectroscopy will reveal the actual motion of electrons in atoms, molecules, and nanoparticles. Clearly, much more work remains to be done in the realm of relativistic plasmas at flux densities exceeding 10^{20} W/cm^2.

It is confirmed and accepted that the soliton communication system can carry the data for more than 40 Gb/s and would be a realistic and efficient optical communication system in near future.

The field of nonlinear optics is alive and has grown much beyond our expectations of four or five decades ago. We believe it will continue to exceed my current expectations in the future.

New Words and Expressions

attosecond 原秒，阿秒
difference-frequency generation (DFG) 差频产生
eigen function 本征函数
femtosecond 飞秒
hypersonic 极超音速的，远超过音速的
nonlinear 非线性的
permittivity [电] 介电常数；电容率
photon echoes 光子回波
piezoelectric 压电的
plasmon 等离振子；等离子体激元
quantum well 量子阱
squeezed state 压缩态
sum-frequency generation (SFG) 和频产生
susceptibility 极化率；磁化系数
transient-response 瞬态响应
tuned 调谐的，已调谐的

Translation

1. Typically, only laser light is sufficiently intense to modify the optical properties of a

material system.

翻译： 通常，只有激光强度足以改变材料系统的光学特性。

2. Nonlinear effects can take place in solids, liquids, gases, and plasmas, and may involve one or more electromagnetic fields as well as internal excitations of the medium.

翻译： 非线性效应可以发生在固体、液体、气体和等离子体中，可能涉及一个或多个电磁场以及介质的内部激发。

3. As a result, there is no coupling between different light beams or between different monochromatic components when they pass through a medium.

翻译： 因此，当不同的光束通过介质时，不同的光束之间或不同的单色分量之间没有耦合。

4. Shortly after the demonstration of the first laser device (a pulsed ruby laser) in 1960, it was found that these simple linear assumptions or conclusions described above were no longer adequate for circumstances in which an intense laser beam was incident on certain types of optical media.

翻译： 在1960年展示了第一台激光设备（脉冲红宝石激光器）后不久，发现上述简单的线性假设或结论已不适用于强激光束入射到某些类型的光学介质上的情况。

5. In general, the contents of nonlinear optics are much more extensive than that of linear optics and, accordingly, the theories of the former are more complicated than that of the latter.

翻译： 一般来说，非线性光学的内容要比线性光学广泛得多，因此非线性光学的理论也比线性光学复杂得多。

6. These experimental demonstrations not only verified the validity of nonlinear polarization theories but also provided an alternative approach to generate coherent optical radiation.

翻译： 这些实验结果不仅验证了非线性偏振理论的有效性，而且为相干光辐射的产生提供了一种新的途径。

7. Another major subject of nonlinear optics is related to the refractive-index change induced by an intense laser beam as well as the impact of this change on the laser beam itself.

翻译： 非线性光学的另一个主要课题与强激光束引起的折射率变化以及这种变化对激光束本身的影响有关。

8. Some of them are the optical analog of the corresponding effects in nuclear magnetic resonance studies.

翻译： 其中一些是核磁共振研究中相应效应的光学模拟物。

9. In accordance with the development of the optical communications, different optical devices with ultrafast response have been designed and fabricated, including the optical switching, optical modulator and detectors.

翻译： 随着光通信技术的发展，人们设计并制作了各种具有超快响应的光学器件，包括光开关、光调制器和探测器。

10. During this period, there are still many research work and progresses on the pulse

compression，the quantum optics，the time-resolved spectroscopic，etc..

翻译： 在此期间，在脉冲压缩、量子光学、时间分辨光谱等方面仍有许多研究工作和进展。

11. It is well known that the brightness or photon degeneracy of a weak optical signal can be dramatically increased based on the coherent amplification through a lasing medium，a stimulated scattering medium，or an optical parametric amplifier system.

翻译： 众所周知，基于通过激光介质，受激散射介质或光学参量放大器系统的相干放大，可以显著提高弱光信号的亮度或光子简并性。

12. The key issue of semi-classical theory in nonlinear optics is to give the expressions of macroscopic nonlinear electric polarization for optical media.

翻译： 非线性光学半经典理论的核心问题是给出光学介质宏观非线性极化的表达式。

Unit 8
Photonic Crystal Fibers

8.1 The Origins of PCFs

Optical fibers are widely used in optical fiber telecommunications, which permits transmission over longer distances and at higher bandwidths (data rates) than other forms of communications. Fibers are used instead of metal wires because signals travel along them with less loss, and they are also immune to electromagnetic interference. Optical fibers are also used for sensors and other applications. Photonic crystals are periodic optical nanostructures that are designed to affect the motion of photons in a similar way that periodicity of a semiconductor crystal affects the motion of electrons. PCF is a new class of optical fiber based on the properties of photonic crystals.

8.1.1 Conventional Optical Fibers

Optical fibers, which rely on total internal reflection (TIR) to guide light and transmit information in the form of short optical pulses over long distances at exceptionally high speeds, are one of the major technological successes of the 20th century. This technology has developed at an incredible rate, from the first low-loss single-mode waveguides in 1970 to being key components of the sophisticated global telecommunication network. Charles K. Kao, widely regarded as the "father of fiber optics" or "father of fiberoptic communications", was awarded half of the 2009 Nobel Prize in Physics for "groundbreaking achievements concerning the transmission of light in fibers for optical communication". Optical fibers have also non-telecom applications, for example, in beam delivery for medicine, machining and diagnostics, sensing, and a lot of other fields. Modern optical fibers represent a careful trade-off between optical losses, optical nonlinearity, group velocity dispersion, and polarization effects. After 30 years of intensive research, incremental steps have refined the capabilities of the system and the fabrication technology nearly as far as they can go.

8.1.2 Photonic Crystal

Photonic crystals occur in nature and in various forms have been studied scientifically for the last 100 years. The idea of photonic crystal originated in 1987 from work in the inhibition of spontaneous emission predicted by E. Yablonovitch and in the field of strong localization of light predicted by S. John. It was subsequently shown that in periodic arrangements of ideally lossless dielectrics, the propagation of light can be totally suppressed at certain wavelengths regardless of propagation direction and polarization. The inhibition does not result from absorption but rather from the periodicity of the arrangement, and is possible (the so-called photonic bandgap, PBG), the density of possible states for the light vanishes, so that even spontaneous emission becomes impossible. Such periodic arrangements of dielectrics have been called photonic crystals, or photonic bandgap materials.

(1) One-dimensional photonic crystals

The simplest device using the principles of photonic crystals is the one-dimensional photonic crystal, well known under the name of the Bragg mirror or the multilayer reflector. It consists of a periodic stack of two alternating dielectric layers. Light propagating in a direction normal to the layers undergoes successive reflection and transmission at each interface between adjacent layers. With an appropriate choice of layer thickness and refractive indices, waves reflected from each interface are in phase, whereas waves transmitted are out of phase. In that case, the transmitted wave components cancel each other out, and only the interference of the reflected components is constructive; the light is totally reflected. This works for a range of wavelengths. Bragg mirrors have been in use for decades, but it is only recently that they have come to be regarded as a special case of photonic crystals. The classical way of analyzing Bragg mirrors with a finite number of layers, uses reflection and transmission matrices for each layer, and it is then quite straightforward to prove through recurrence relationships that reflection call is perfect with an infinite number of layers. There is nevertheless another approach to deal with a stack having an infinite number of layers, originating from solid state physics. If the stack is infinite, it has a discrete translational symmetry. The Bloch theorem then applies, and solutions to the propagation equation in the stack are Bloch waves. Hence two wave vectors differing by a vector of the reciprocal lattice associated with the periodic stacking are physically the same: the dispersion diagram "folds back" along the limits of the Brillouin zone. At the edge of the Brillouin zone, two solutions exist having the same wave vector but different frequencies, and in between those two frequencies no solutions exist at all. The gap of frequencies for which no solutions exist is called a photonic bandgap. Note that, reflection from Bragg mirrors was thought to be possible only within a relatively narrow range of angles of incidence.

(2) Two and three-dimensional photonic crystals

Photonic crystals with two or three-dimensional periodicity can be seen as a generalization of Bragg mirrors. The simple approach with reflection and transmission matrices cannot be applied analytically here, and this is probably why their properties were discovered relatively recently, although, for example, important work on stacked grids for filtering in the far infrared was carried out by R. Ulrich in the 1960s. The Bloch approach can be used similarly, and shows that bandgaps can still open up. The point of using periodicities along two or three-dimensions is to open up an omnidirectional bandgap: for the Bragg mirror, bandgaps usually only exist for a narrow range of angles of incidence, and propagation parallel to the Bragg, layers can never be inhibited. With photonic crystals having a two-dimensionally periodic arrangement of parallel rods, bandgaps can exist for all directions of propagation in the plane of periodicity, and photonic crystals with three-dimensional periodicity propagation of light in all directions can be prohibited. When a bandgap exists regardless of direction of propagation and polarization, one speaks of a total photonic bandgap.

Photonic crystals with two-dimensional periodic arrangements are usually either made of parallel dielectric (or metallic) rods in air, or through drilling or etching holes in a dielectric material. In the field of integrated optics, holes of a fraction of a micrometer etched in slab waveguides are very promising for integrated photonic circuits, and have been successfully demonstrated experimentally. Photonic crystals with three-dimensional periodicity are a bit trickier to achieve. Yablonovitch suggested drilling an array of holes at three different angles into a dielectric material. The so-called wood-pile structure has attracted much attention and recent progress with artificial inverse opals is promising.

Note that the term photonic crystal was originally introduced to refer to photonic bandgap. It seems that it is now progressively more often used to any kind of periodic arrangement of dielectrics or materials, with or without photonic bandgaps. The latter generalization of the term makes sense considering that in solid state physics, a crystal is named so on account of the periodicity of its lattice, with bandgaps appearing in certain case. Usual practice is then to reserve the term photonic bandgap material for a photonic crystal having a photonic bandgap.

The interest of researchers and engineers in several laboratories, since the 1980s, has been attracted by the ability to structure materials on the scale of the optical wavelength, a fraction of micrometers or less, in order to develop new optical medium, known as photonic crystals. Photonic crystals rely on a regular morphological microstructure, incorporated into the material, which radically alters its optical properties. They represent the extension of the results obtained for semiconductors into optics. In fact, the band structure of semiconductors is the outcome of the interactions between electrons and the periodic variations in potential created by the crystal lattice. By solving the Schrodinger's wave equation for a periodic potential, electron energy states separated by forbidden bands are obtained. PBGs can be obtained in photonic crystals, where periodic variations in dielectric constant that is in refractive index substitute variations in electric potential, as well as the classical wave equation for the magnetic field replaces the Schrodinger's equation.

8. 1. 3 From Conventional Optical Fibers to PCFs

In 1991, Philip Russell, who was interested in Yablonovitch's research, got his big "crazy" idea for "something different", during CLEO/QELS conference. Russell's idea was that light could be trapped inside a fiber hollow core by creating a two-dimensional photonic crystal in the cladding that is a periodic wavelength-scale lattice of microscopic air-holes in the glass. The basic principle is the same which is the origin of the color in butterfly wings and peacock feathers, that is all wavelength-scale periodic structures exhibit ranges of angle and color, stop bands, where incident light is strongly reflected. When properly designed, the photonic crystal cladding running along the entire fiber length can prevent the escape of light from the hollow core. These new fibers are called PCFs, since they rely on the unusual properties of photonic crystals.

8.2 The History of PCFs

The first fiber with a photonic crystal structure was reported by Russell and his colleagues in 1996. Even if it was a very interesting research development, the first PCF did not have a hollow core, and, consequently, it did not rely on a photonic bandgap for optical confinement. In fact, in 1996 Russell's group at University of Southampton could produce fiber with the necessary air-hole triangular lattice, but the air-holes were too small to achieve a large air-filling fraction, which is fundamental to realize a PBG. Measurements have shown that this solid core fiber formed a single-mode waveguide that only the fundamental mode was transmitted, over a wide wavelength range. Moreover, the first PCF had very low intrinsic losses, due to the absence of doping elements in the core, and a silica core with an area about ten times larger than that of a conventional single-mode fiber (SMF), thus permitting a corresponding increase in optical power levels. In 1997 Birks reported an endlessly single-mode (ESM) PCF, which is developed by embedding a central core in a two-dimensional photonic crystal with a micrometer-spaced hexagonal array of air holes. Such a fiber can be single mode for any wavelength. It is shown that its useful single-mode range is bounded by a bend-loss edge at short wavelength as well at long wavelengths. A high-speed optical-transmission experiment was successfully demonstrated in an ultra low-loss polarization maintaining PCF by K. Suzuki in 2003. The fiber loss and modal birefringence at 1550 nm were 1.3 dB/km and 1.4×10^{-3}, respectively. A 10 Gb/s bi-directional optical signal was successfully transmitted through the 1.5 km fiber.

After moving his research group to the University of Bath in 1996, where PCF fabrication techniques were steadily refined, Russell and his co-workers were able to report, in 1999, the first single-mode hollow core fiber, in which confinement was due by a full two-dimensional PBG. They realized that the photonic bandgap guiding mechanism is very robust, since light remains well confined in the hollow core, even if tight bends are formed in the fiber. However, it is highly sensitive to small fluctuations in the fiber geometry, for example, to variations in the air-hole size.

Initial production techniques were directed simply at the task of making relatively short lengths of fiber in order to do the basic science, but many research teams are now working hard to optimize their PCF production techniques, in order to increase the lengths and to reduce the losses.

8.3 Guiding Light in PCFs

In order to form a guided mode in an optical fiber, it is necessary to introduce light into the core with a value of β that is the component of the propagation constant along the fiber axis, which cannot propagate in the cladding. The highest β value that can exist in an infinite homogeneous medium with refractive index n is $\beta = nk_0$, being k_0 the free-space propagation constant. All the smaller values of β are allowed. A two-dimensional photonic crystal, like any other material, is characterized by a maximum value of β which can propagate. At a particular wavelength, this corresponds to the fundamental mode of an infinite slab of the material, and this β value defines the elective refractive index of the material.

8. 3. 1 Modified Total Internal Reflection

It is possible to use a two-dimensional photonic crystal as a fiber cladding, by choosing a core material with a higher refractive index than the cladding effective index. An example of this kind of structures is the PCF with a silica solid core surrounded by a photonic crystal cladding with a triangular lattice of air-holes. These fibers, also known as index-guiding PCFs, guide light through a form of TIR, called modified TIR. However, they have many different properties with respect to conventional optical fibers.

8. 3. 2 Photonic Bandgap Guidance

Optical fiber designs completely different from the traditional ones result from the fact that the photonic crystal cladding have gaps in the ranges of the supported modal index β/k_0 where there are no propagating modes. These are the PBGs of the crystal, which are similar to the two-dimensional bandgaps which characterize planar light wave circuits, but in this case they have propagation with a non zero value of β. It is important to underline that gaps can appear for values of modal index both greater and smaller than unity, enabling the formation of hollow core fibers with bandgap material as a cladding. These fibers, which cannot be made using conventional optics, are related to Bragg fibers, since they do not rely on TIR to guide light. In fact, in order to guide light by TIR, a lower-index cladding material surrounding the core is necessary, but there are no suitable low-loss materials with a refractive index lower than air at optical frequencies. The first PGF which exploited the PBG effect to guide light was reported in 1998. Notice that its core is formed by an additional air-hole in a honeycomb lattice. This PCF could only guide light in silica that is in the higher-index material.

Hollow core guidance had to wait until 1999, when the PCF fabrication technology had advanced to the point where larger air-filling fractions, required to achieve a PBG for air-guiding, became possible. Notice that an air guided mode must have $\beta/k_0 < 1$, since this condition guarantees that light is free to propagate and form a mode within the hollow core, while being unable to escape into the cladding. The first hollow core PCF had a simple triangular lattice of air-holes, and the core was formed by removing seven capillaries in the center of the fiber cross-section. By producing a relatively large core, the chances of finding a guided mode were improved. When white light is launched into the fiber core, colored modes are transmitted, thus indicating that light guiding exists only in restricted wavelength ranges, which coincide with the photonic bandgaps.

8.4 The Properties of PCFs

Due to the huge variety of air-holes arrangements, PCFs offer a wide possibility to control the refractive index contrast between the core and the photonic crystal cladding and, as a consequence, novel and unique optical properties.

Since PCFs can be divided into two modes of operation, according to their mechanism for confinement. Those with a solid core or a core with a higher average index than the photonic crystal cladding, can operate on the same index-guiding principle as conventional optical fiber, however, they can have a much higher effective-index contrast between core and cladding, and therefore can have much stronger confinement for applications in nonlinear optical devices, polarization-maintaining fibers. Alternatively, one can create a photonic bandgap fiber, in which the light is confined by a photonic bandgap created by the photonic crystal cladding, such a bandgap, properly designed, can confine light in a lower-index core and even a hollow (air) core. Bandgap fibers with hollow cores can potentially circumvent limits imposed by available materials, for example to create fibers that guide light in wavelengths for which transparent materials are not available (because the light is primarily in the air, not in the solid materials). Another potential advantage of a hollow core is that one can dynamically introduce materials into the core, such as a gas that is to be analyzed for the presence of some substance.

8.4.1 Solid Core PCFs

Index-guiding PCFs, with a solid glass region within a lattice of air-holes, offer a lot of new opportunities, not only for applications related to fundamental fiber optics. These opportunities are related to some special properties of the photonic crystal cladding, which are due to the large refractive index contrast and the two-dimensional nature of the microstructure, thus affecting the mode property, the dispersion, the smallest attainable core size, the number of guided modes and the numerical aperture and the birefringence.

(1) Endlessly single-mode property

As already stated, the first solid core PCF, which consisted of a triangular lattice of air-holes with a diameter d of about 300 nm and a hole-to-hole spacing A of 2.3 μm, did not ever seem to become multi-mode in the experiments, even for short wavelengths. In fact, the guided mode always had a single strong central lobe filling the core.

Russell has explained that this particular ESM behavior can be understood by viewing the air-hole lattice as a modal filter or "sieve". Since light is evanescent in air, the airholes act like strong barriers, so they are the "wire mesh" of the sieve. The field of the fundamental mode, which fits into the silica core with a single lobe of diameter between zeros slightly equal to 2Λ, is the "grain of rice" which cannot escape through the wire mesh, being the silica gaps between the air-holes belonging to the first ring around the core too narrow. On the contrary, the lobe dimensions for the higher-order modes are smaller, so they can slip between the gaps. When the ratio d/Λ, that is the air-filling fraction of the photonic crystal cladding, increases, successive higher-order modes become trapped. A proper geometry design of the fiber cross-section thus guarantees that only the fundamental mode is guided. More detailed studies of the properties of triangular PCFs have shown that this occurs for $\Lambda < 0.4$.

By exploiting this property, it is possible to design very large-mode area fibers, which can be successfully employed for high-power delivery, amplifiers, and lasers. Moreover, by doping the core in order to slightly reduce its refractive index, light guiding can be turned off completely at wavelengths shorter than a certain threshold value.

(2) Dispersion tailoring

The tendency for different light wavelengths to travel at different speeds is a crucial factor in the telecommunication system design. A sequence of short light pulses carries the digitized information. Each of these is formed from a spread of wavelengths and, as a result of chromatic dispersion, it broadens as it travels, thus obscuring the signal. The magnitude of the dispersion changes with the wavelength, passing through zero at 1.3 μm in conventional optical fibers.

In PCFs, the dispersion can be controlled and tailored with unprecedented freedom. In fact, due to the high refractive index difference between silica and air, and to the flexibility of changing air-hole sizes and patterns, a much broader range of dispersion behaviors can be obtained with PCFs than with standard fibers.

On the contrary, very flat dispersion curves can be obtained in certain wavelength ranges in PCFs with small air-holes, which is with low air-filling fraction.

(3) High nonlinearity

An attractive property of solid core PCFs is that effective index contrasts much higher than in conventional optical fibers can be obtained by making large air-holes, or by reducing the core dimension, so that the light is forced into the silica core. The nonlinear coefficient $\gamma(\lambda)$ of PCFs can be expressed as

$$\gamma(\lambda) = \frac{2\pi n_2}{\lambda A_{\text{eff}}} \tag{8.1}$$

where A_{eff} is model effective area, and n_2 is the nonlinear index of silica. In this way a strong confinement of the guided-mode can be reached, thus leading to enhanced nonlinear effects, due to the high field intensity in the core. Moreover, a lot of nonlinear experiments require specific dispersion properties of the fibers. As a consequence, PCFs can be successfully exploited to realize nonlinear fiber devices, with a proper dispersion, and this is presently one of their most important applications.

(4) Large-mode area

By changing the geometric characteristics of the fiber cross-section, it is possible to design PCFs with completely different property, that is with large effective area. The typical cross-section of this kind of fibers, called large mode area (LMA) PCFs, consists of a triangular lattice of air-holes where the core is defined by a missing air-hole. In this condition it is the core size or the pitch that determines the zero-dispersion wavelength λ_0, the mode field diameter (MFD) and the numerical aperture (NA) of the fiber. LMA PCFs are usually exploited for high-power applications, since fiber damage and nonlinear limitations are drastically reduced. In particular, LMA fibers are currently used for applications at short wavelengths, that is in ultraviolet (UV) and visible bands, like the generation and delivery of high-power optical beams for laser welding and machining, optical lasers, and amplifiers, providing significant advantages with respect to traditional optical fibers.

Conventional active fibers for lasers and amplifiers are basically standard transmission fibers whose core region has been doped with rare earth elements. These fibers, also known as "core-pumped", are usually pumped with single mode pump lasers. Due to its power limitations, this kind of fiber is unsuitable for high-power applications, on the order of 1 W, and upwards. High-power fibers are usually designed with a double-cladding structure, where a second low-index region acts as a cladding for a large pump core.

(5) High birefringence

Birefringent fibers, where the two orthogonally polarized modes carried in a singlemode fiber propagate at different rates, are used to maintain polarization states in optical devices and subsystems. The guided modes become birefringent if the core microstructure is deliberately made twofold symmetric, for example, by introducing capillaries with different wall thicknesses above and below the core. By slightly changing the air-hole geometry, it is possible to produce levels of birefringence that exceed the performance of conventional birefringent fiber by an order of magnitude. It is important to underline that, unlike traditional polarization maintaining fibers, such as bow tie, elliptical-core or Panda, which contain at least two different glasses, each one with a different thermal expansion coefficient, the birefringence obtainable with PCFs is highly insensitive to temperature, which is an important feature in many applications.

(6) High numerical aperture

Highly numerical aperture (HNA) PCFs have a central part surrounded by a ring of relatively large air holes. HNA-PCFs have the importance of minimizing the width of silica bridges in order to obtain a low cladding index. Several such fibers have been fabricated and the measured properties as a function of wavelength and web thickness follow the predictions well. These fibers show the highest NAs reported of 0.88 over a 41 m length at a wavelength of 1.1 μm, rising to NA at 1.54 μm, and decreasing to NA 0.65 at 450 nm. Such structures will lead to performance improvements for cladding-pumped lasers and increased sensitivity in collection of incoherent light.

8.4.2 Hollow Core PCFs

Hollow core PCFs have great potential, since they exhibit low nonlinearity and high damage threshold, thanks to the air-guiding in the hollow core and the resulting small overlap between silica and the propagating mode. As a consequence, they are good candidates for future telecommunication transmission systems.

Another application, perhaps closer to fruition, which can successfully exploit these advantages, offered by air-guiding PCFs, is the delivery of high-power continuous wave (CW), nanosecond and sub-picosecond laser beams, which are useful for marking, machining and welding, laser-Doppler velocimetry, laser surgery, and THz generation. In fact, optical fibers would be the most suitable delivery means for many applications, but at present they are unusable, due to the fiber damage and the negative nonlinear effects caused by the high optical powers and energies, as well as to the fiber group-velocity dispersion, which disperses the short pulses. These limitations can be substantially relieved by considering hollow core fibers.

Moreover, air-guiding PCFs are suitable for nonlinear optical processes in gases, which require high intensities at low power, long interaction lengths and good-quality transverse, beam profiles. For example, it has been demonstrated that the threshold for stimulated Raman scattering in hollow core fibers filled with hydrogen is orders of magnitude below that obtained in previous experiments. In a similar way, PCFs with a hollow core can be used for trace gas detection or monitoring, or as gain cells for gas lasers. Finally, the delivery of solid particles down a fiber by using optical radiation pressure has been demonstrated. In particular, only 80 mW of a 514 nm argon laser light was enough to levitate and guide 5 μm polystyrene spheres along a 15 cm length of PCF with a hollow core diameter of 20 μm.

8.5 The Fabrication of PCFs

One of the most important aspects in designing and developing new fibers is their fabrication process. PCFs have been realized by "introducing" air-holes in a solid glass material. This has several advantages, since air is mechanically and thermally compatible with most materials, it is transparent over a broad spectral range, and it has a very low refractive index at optical frequencies. Fibers fabricated using silica and air have been accurately analyzed, partly because most conventional optical fibers are produced from fused silica. This is also an excellent material to work with, because viscosity does not change much with temperature and it is relatively cheap. Moreover, filling the holes of a silica-air structure opens up a wide range of interesting possibilities, such as the bandgap guidance in a low-index core made of silica when the holes are filled with a high-index liquid.

The traditional way of manufacturing optical fibers usually involves two main steps: fabrication a fiber preform and drawing it with a high-temperature furnace in a tower setup. For conventional silica-based optical fibers, both techniques are very mature. The different vapor deposition techniques, for example, the modified chemical vapor deposition (MCVD), the vapor axial deposition (VAD), and the outside vapour deposition

(OVD), are all tailored for the fabrication of circular symmetric fiber preforms. Thus, the deposition can be controlled in a very accurate way only in the radial direction without significant modifications of the methods. Moreover, producing conventional single-mode optical fibers requires core and cladding materials with similar refractive index values, which typically differ by around 1%, and are usually obtained by vapor deposition techniques. On the contrary, designing PCFs requires a far higher refractive index contrast, differing by perhaps 50% ~ 100%. As a consequence, all the techniques previously described are not directly applicable to the fabrication of preforms for PCFs, whose structure is not characterized by a circular symmetry.

Differently from the drawing process of conventional optical fibers, where viscosity is the only important material parameter, several forces are important in the case of PCFs, such as viscosity, gravity, and surface tension. This is due to the much larger surface area in a microstructured geometry, and to the fact that many of the surfaces are close to the fiber core, thus making surface tension relatively much more important. As a consequence, the choice of the base material strongly influences the technological issues and applications in the PCF fabrication process.

8. 5. 1 Stack-and-draw Technique

In order to fabricate a PCF, it is necessary, first, to create a preform, which contains the structure of interest, but on a macroscopic scale. One possibility to exploit for the PCF fabrication is the drilling of several tens to hundreds of holes in a periodic arrangement into one final preform. However, a different and relatively simple method, called stack-and-draw, introduced by Birks et al. in 1996, has become the preferred fabrication technique in the last years, since it allows relatively fast, clean, low-cost, and flexible preform manufacture.

The solid core PCF preform is realized by stacking by hand a number of capillary silica tabes and solid silica rods to form the desired air-silica structure. This way of realizing the preform allows a high level of design flexibility, since the core size and shape, as well as the index profile throughout the cladding region can be controlled. After the stacking process, the capillaries and rods are held together by thin wires and fused together during an intermediate drawing process, where the preform is drawn into preform canes. This intermediate step is important in order to provide numerous preform canes for the development and optimization of the later drawing of the PCFs to their final dimensions. Then, the preform is drawn down on a conventional fiber drawing tower, greatly extending its length, while reducing its cross-section, from a diameter of 20 mm to 80 ~200 μm, as shown in Figure 8-1. With respect to standard optical fibers, which are usually drawn at temperatures around 2100 ℃, a lower temperature level, that is 1900 ℃, is kept during the PCF drawing since the surface tension can otherwise lead to the air-hole collapse. In order to carefully control the air-hole size during the drawing process, it is useful to apply to the inside of the preform a slight overpressure relative to the surroundings, and to properly adjust the drawing speed. In summary, time dynamics, temperature, and pressure variations are all significant parameters which should be accurately controlled during the PCF fabrication. Finally, the PCFs are coated to provide a protective standard jacket, which allows the robust handling of the fibers. The final PCFs are comparable to standard fibers in both robustness and physical dimensions, and can be both striped and cleaved using standard tools.

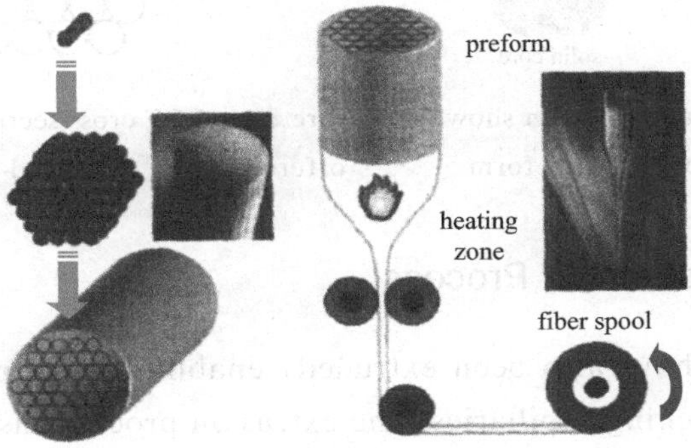

Figure 8-1 Scheme of the PCF fabrication process

The fabrication process of hollow core PCF preform is similar to the solid core PCF preform fabrication process. The different procedure is displacing rods and keeping air core, represented in Figure 8-2. It is important to underline that the stack-and-draw procedure, represented in Figure 8-3, proved highly versatile, allowing complex lattices to be assembled from individual stackable units of the correct size and shape. Solid, empty, or doped glass regions can be easily incorporated, as described in Figure 8-1. A wide range of different structures have been made by exploiting this technique, each with different optical properties. Moreover, overall collapse ratios as large as about 50000 times have been realized, and continuous holes as small as 25 nm in diameter have been demonstrated, earning an entry in the Guinness Book of Records in 1999 for the World's Longest Holes. A very important issue is the comparison of the PCF stack-and-draw procedure with the vapor deposition methods usually employed for standard optical fibers. Obviously, it is more difficult that the preforms for conventional optical fibers become contaminated, since their surface area is smaller. Moreover, the stacking method requires a very careful handling, and the control of air-hole dimensions, positions, and shapes in PCFs makes the drawing significantly more complex. Finally, it is important to underline that the fabrication process of PCFs with a hollow core, realized by removing some elements from the stack center, is much more difficult than that of standard optical fibers, even if at present fibers with low loss and practical lengths have been obtained.

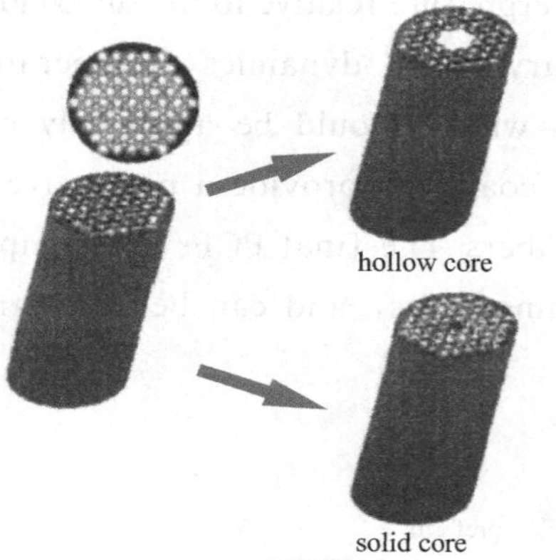

Figure 8-2 Schematic diagram shown the fabrication of PCFs preform

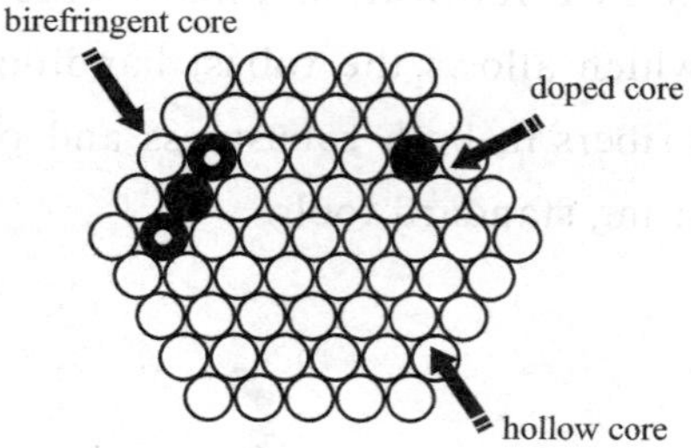

Figure 8-3 PCF cross-section, showing the flexibility offered by the stack-and-draw fabrication process

8.5.2 Extrusion Fabrication Process

Silica-air preforms have also been extruded, enabling the formation of structures not readily attainable by stacking capillaries. The extrusion process has been recently applied to other glasses, which are not as readily available in tube form as silica, like compound glasses. These materials, which provide a lot of interesting properties, like an extended wavelength range for transmission and higher values of the nonlinear coefficient, can be used to fabricate preforms through the extrusion process due to their lower softening temperatures, which make the fabrication procedure easier.

In this fabrication process, a molten glass is forced through a die containing a suitably designed pattern of holes. Extrusion allows fiber to be drawn directly from bulk glass, using a fiber-drawing tower, and almost any structure, crystalline or amorphous, can be produced. It works for many materials, including chalcogenides, polymers, and compound glasses. However, selective doping of specified regions, in order to introduce rare earth ions or render the glass photosensitive, is much more difficult. Different PCFs produced by the extrusion process have been presented in literature. In particular, the fabrication of the first non-silica glass PCF by exploiting this technique has been reported in 2002 by Kiang et al. A commercial glass, called Schott SF57 glass, has been used, which has a softening temperature of only 519 ℃ and a high lead concentration, which causes a relatively high refractive index of 1.83 at a wavelength of 633 nm and of 1.80 at 1530 nm. This material is interesting since its nonlinear refractive index is 4.1×10^{-19} W^2/m at 1060 nm, which is more than one order of magnitude larger than that of pure silica. Another highly nonlinear PCF has been fabricated with the bismuth-oxide-based glass, which has proved to be an attractive novel material for nonlinear devices and compact Er^{3+}-doped amplifiers. The fiber fabrication presented in consists of three steps. In the first step, the structured preform of 16 mm outer diameter and the jacket tube are extruded. In the second step, the preform is reduced in scale on a fiber-drawing tower to a cane of about 1.6 mm diameter. In the last step, the cane is inserted within the jacket tube, and this assembly is drawn down to the final fiber.

8.6 The Applications of PCFs

The diversity of new or improved features, beyond what conventional fiber offers, means that PCF is finding an increasing number of applications in ever-widening areas of science and technology. Let us sample a few of the more intriguing and important ones.

8.6.1 High Power and Energy Transmission

Larger mode areas allow higher power to be carried before the onset of intensity related nonlinearities or damage, with obvious benefits for delivery of high laser power, and for high-power fiber amplifiers and lasers. ESM-PCF's ability to remain single mode at all wavelengths where it guides, and for all scales of structure, means that the core area can be increased without the penalty of introducing higher order guided modes. This also suggests that it should have superior power-handling properties, with applications in, for example, the field of laser machining. A key issue is bend loss, and as we have seen, it turns out that PCF offers a wider bandwidth of useful single-mode guidance than high-delta SMF, because it can operate in the multimode regime of SMF while remaining single mode. This also shifts the long wavelength bend edge to longer wavelengths than is possible in standard fibers.

Hollow core PCF is also an excellent candidate for transmitting high continuous-wave power as well as ultra-short pulses with very high peak powers. Solitons have been reported at 800 nm using a Ti: sapphire laser and at 1550 nm with durations of 100 fs and peak powers of 2 MW. The soliton energy is of course determined by the effective value of and the magnitude of the anomalous GVD. The GVD changes sign across the bandgap, permitting choice of normal or anomalous dispersion depending upon the application. Further studies have explored the ultimate power handling capacity of hollow core PCF.

8.6.2 Fiber Lasers and Amplifiers

PCF lasers can be straightforwardly produced by incorporating a rare-earth-doped cane in the preform stack. Many different designs can be realized, such as cores with ultra-large mode areas for high power, and structures with multiple lasing cores. Cladding-pumping geometries for ultrahigh power can be fashioned by incorporating a second core (much larger and multimode) around a large off-center ESM lasing core. Using microstructuring techniques, this "inner cladding waveguide" can be suspended by connecting it to an outer glass tube with very thin webs of glass. This results in a very large effective index step and thus a high-numerical aperture (>0.9), making it easy to launch and guide light from high-power diode-bar pump lasers. The multimode pump light is efficiently absorbed by the lasing core, and high-power single-mode operation can be achieved. Microchip-laser seeded Yb^{3+}-doped PCF amplifiers, generating diffraction-limited 0.45 ns duration pulses with a peak power of 1.1 MW and a peak spectral brightness of greater than 10 kW/ (cm^2 · sr · Hz), have been reported.

Hollow core PCF with its superior power-handling and designable GVD is ideal as the last compression stage in chirped-pulse amplification schemes. This permits operation at power densities that would destroy conventional glass-core fibers.

8.6.3 Gas-based Nonlinear Optics

A longstanding challenge in photonics is how to maximize nonlinear interactions between laser light and low-density media such as gases. Efficient nonlinear processes require high intensities at low power, long interaction lengths, and good-quality transverse beam profiles. No existing solution comes close to the performance offered by hollow core PCF. At a bore diameter of 10 m, for example, a focused free-space laser beam is marginally preferable to a capillary, whereas a hollow-core PCF with 13 dB/km attenuation is 105 times more effective. Such enhancements are rare in physics and point the way to improvements in all sorts of nonlinear laser-gas interactions. Discussed next are just two examples from a rich prospect of enhanced, and more practical, ultra-low-threshold gas-based nonlinear optical devices.

An example is ultra-low-threshold stimulated Raman scattering in molecular gases. Raman scattering is caused by molecular vibrations, typically in the multi-THz range, that interact spontaneously with the laser light, shifting its frequency both up (anti-Stokes) and down (Stokes) in two separate three-wave parametric interactions. At high intensities, the Stokes wave becomes strong and beats with the pump laser light, driving the molecular oscillations more strongly. This further enhances the Stokes signal, so that ultimately, above a certain threshold power, the major fraction of the pump power is converted to the Stokes frequency. The energy lost to molecular vibrations is dissipated as heat. A stimulated Raman threshold was recently observed in a hydrogen filled hollow core PCF at pulse energies 100 times lower than previously possible. Another field where hollow-core fiber is likely to have a major impact is that of high harmonic generation. When gases such as argon are subjected to ultra-short (few fs) high-energy (few mJ) pulses, usually from a Ti: sapphire laser system operating at 800 nm wavelength, the extremely high, short duration electric field momentarily ionizes the atoms, and very high harmonics of the laser frequency are generated during the recombination process. Ultraviolet and even X-ray radiation can be produced in this way. It is tantalizing to speculate that hollow core PCF could bring this process within the reach of compact diode-pumped laser systems, potentially leading to table-top X-ray sources for medicine, lithography, and X-ray diagnostics.

8.6.4 Supercontinuum Generation

PCFs with extremely small solid glass cores and very high air-filling fractions not only display unusual chromatic dispersion but also yield very high optical intensities per unit power. Thus one of the most successful applications of PCF is nonlinear optics, where high effective nonlinearities, together with excellent control of chromatic dispersion, are essential for efficient devices.

A dramatic example is supercontinuum generation. Supercontinuum generation is the formation of broad continuous spectra by propagation of high power pulses through nonlinear media. When ultra-short, high-energy pulses travel through a material, their frequency spectrum can experience giant broadening due to a range of interconnected nonlinear effects. Until recently this required a regeneratively amplified Ti-sapphire laser operating at 800 nm wavelength. Pulses from the master oscillator (100 MHz repetition rate, 100 fs duration, few nJ energy) are regeneratively amplified up to 1 mJ. Because the amplifier needs to be recharged between pulses, the repetition rate is only around 1 kHz. Thus, there was great excitement when it was discovered that highly nonlinear PCF, designed with zero chromatic dispersion close to 800 nm, displays giant spectral broadening when the 100 MHz pulse train from the master oscillator is launched into just a few cm of fiber. The pulses emerge from a tiny aperture and last only a few ps. They have the bandwidth of sunlight but are 10 times brighter. Not surprisingly, this source is finding many uses, e. g., in optical coherence tomography. The huge bandwidth and high spectral brightness of the supercontinuum source make it ideal for all sorts of spectroscopy. Measurements that used to take hours and involve counting individual photons can now be made in a fraction of a second. Furthermore, because the light emerges from a microscopic aperture, it is uniquely easy to perform spectroscopy with very high spatial resolution.

8. 6. 5 Telecommunications

There are many potential applications of PCF or PCF-based structures in telecommunications, although whether these will be adopted remains an open question. One application that seems quite close to being implemented is the use of solid core PCF or "hole-assisted" SMF for fiber-to-the-home, where the lower bend-loss is the attractive additional advantage offered by the holey structure. Other possibilities include dispersion-compensating fiber and hollow core PCF for long-haul transmission. Additional opportunities exist in producing bright sources of correlated photon pairs for quantum cryptography, parametric amplifiers with improved characteristics, highly nonlinear fiber for all-optical switching and amplification, acetylene-filled hollow core PCF for frequency stabilization at 1550 nm, and the use of sliced SC spectra as wavelength division multiplexing (WDM) channels. There are also many possibilities for ultra-stable in-line devices based on permanent morphological changes in the local holey structure, induced by heating, collapse, stretching, or inflation. The best reported loss in solid core PCF, from a group in Japan, stands at 0. 28 dB/km at 1550 nm, with a Rayleigh scattering coefficient of 0. 85 dB/(mm • km). A 100 km length of this fiber was used in the first PCF-based penalty-free dispersion-managed soliton transmission system at 10 Gbit/s. The slightly higher attenuation compared with that in

SMF is due to roughness at the glass-air interfaces.

Hollow core PCF is radically different from solid core SMF in many ways. This makes it difficult to predict whether it could be successfully used in long-haul telecommunications as a realistic competitor for SMF-28. The much lower Kerr nonlinearities mean that WDM channels can be much more tightly packed without nonlinear crosstalk, and the higher power-handling characteristics mean that more overall power can be transmitted. The effective absence of bend losses is also a significant advantage, particularly for short-haul applications. On the other hand, work still needs to be done to reduce the losses to 0. 2 dB/km or lower, and to understand and control effects such as polarization mode dispersion, differential group delay, and multipath interference. It is interesting that the low-loss window of a plausible hollow core PCF is centered at 1900 nm, because light travels predominantly in the hollow regions, completely changing the balance between scattering and infrared absorption.

The large glass-air refractive index difference makes it possible to design and fabricate PCFs with high levels of group velocity dispersion. A PCF version of the classical W-profile dispersion-compensating fiber was reported in 2005, offering slope-matched dispersion compensation for SMF-28 fiber at least over the entire C-band (Figure 8-4). The dispersion levels achieved 1200 ps/(nm・km) indicate that only 1 km of fiber is needed to compensate for 80 km of SMF-28. The PCF was made deliberately birefringent to allow control of polarization mode dispersion.

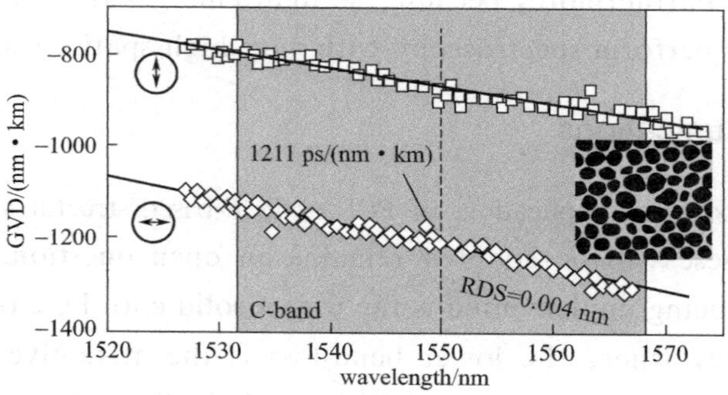

Figure 8-4 Performance of a PCF designed to provide slope-matched dispersion compensation

8. 6. 6 Optical Sensors

Sensing is so far a relatively unexplored area for PCFs, although the opportunities are myriad, spanning many fields including environmental monitoring, biomedical sensing, and structural monitoring. Solid core PCF has been used in hydrostatic pressure sensing. Multicore PCF has been used in bend and shape sensing and Doppler difference velocimetry. Fibers with a central single-mode core surrounded by a highly multimode cladding waveguide are useful in applications such as two-photon fluorescence sensing, where short pulses are delivered to the sample by the central single-mode core, and the resulting multiphoton fluorescence, efficiently collected by inner cladding waveguide, has a large numerical aperture. Sensitivity enhancements of 20 times have been reported. Given the high level of interest, and large amount of effort internationally, it seems very likely that many more important sensing applications of PCF will emerge over the next few years.

8.6.7 Gratings in PCF

Bragg mirrors are normally written into the Ge-doped core of an SMF by UV light, using either two-beam interferometry or a zero-order nulled phase mask. The mechanism for refractive index change is related to the UV Corning SMF-28 over the C-band. Photosensitivity of the glass is quite complex, with several different regimes of operation. In PCF, the presence of many holes in the cladding will inevitably scatter the UV light strongly and reduce (or enhance) the field amplitudes in the core. In addition, pure silica glass is only weakly photosensitive, requiring exposure to very intense light for formation of useful Bragg gratings. There have nevertheless been several reports of the inscription of Bragg gratings into pure-silica PCF, using 125 fs pulses at 800 nm wavelength and multiphoton processes. In the case of polymer PCF, low-power CW laser sources at 325 nm wavelength is sufficient to write Bragg gratings for 1570 nm operation.

Although long-period fiber gratings (LPFGs) and PCFs have emerged at the same time and been around for almost ten years, the fabrications of fiber components in PCFs have attracted great attention in recent years. The potential applications of PCF-LPFG devices for gas sensing have been discussed. Unlike the PCF-based gas sensor that detects the analytes by the interaction of light with gases through the absorption of the evanescent wave in the holes of fiber cladding, the PCF-LPFG gas sensing works by the interrogating of the shifts of different resonance wavelength and strength of core-cladding mode coupling in the transmission spectrum. The advantages of the PCF-LPFG sensing devices are: high temperature insensitive and stability; compactness when packaged; practical use under hazardous conditions and in high temperature environment.

8.7 Future Perspectives

PCFs have flexibility of design and can achieve many extraordinary characteristics, such as novel dispersion properties, high birefringence, single mode guidance over a wide spectral range, high nonlinearity, photonic bandgap guidance and others different from conventional fibers have, Although PCFs have only a near-20-year development history and the applications of PCFs are not diffuse, the developments of PCFs are rapid. As time progresses, there will doubtless be new scientists with new ideas and new experiments which will expand the role of PCFs in fiber optical research.

New Words and Expressions

attenuation 衰减
chalcogenide 硫族（元素）化物，氧属（元素）化物
fluorescence 荧光，荧光性
jacket 护套
large mode area 大模场面积
localization 局域化，局限，定位
mode field diameter 模场直径
numerical aperture 数值孔径
preform 预制棒
stack-and-draw 堆拉法
supercontinuum 超连续谱
vapor axial deposition 汽相轴向沉积
wavelength division multiplexing 波分复用

Translation

1. Optical fibers are widely used in optical fiber telecommunications, which permits transmission over longer distances and at higher bandwidths (data rates) than other forms of communications.

翻译：光纤广泛应用于通信中，与其他形式的通信相比，光纤可在更长的距离和更高的带宽（数据速率）下进行传输。

2. Fibers are used instead of metal wires because signals travel along them with less loss, and they are also immune to electromagnetic interference.

翻译：使用光纤代替金属线是因为信号沿着光纤传输时损耗较小，而且它们也不受电磁干扰。

3. Photonic crystals are periodic optical nanostructures that are designed to affect the motion of photons in a similar way that periodicity of a semiconductor crystal affects the motion of electrons.

翻译：光子晶体是一种周期性的光学纳米结构，它可以被设计成像半导体晶体周期性影响电子运动一样影响光子的运动。

4. This technology has developed at an incredible rate, from the first low-loss single-mode waveguides in 1970 to being key components of the sophisticated global telecommunication network.

翻译：这项技术以惊人的速度发展，从1970年第一个低损耗的单模波导到如今成为复杂的全球电信网络的关键部件。

5. Optical fibers have also non-telecom applications, for example, in beam delivery for medicine, machining and diagnostics, sensing, and a lot of other fields.

翻译：光纤在非电信领域也有许多应用，例如用于医学、机械加工和诊断、传感和许多其他领域的光束传输。

6. Photonic crystals occur in nature and in various forms have been studied scientifically for the last 100 years.

翻译：光子晶体在自然界中存在，在过去的100年里，人们一直在对它的各种形式进行科学研究。

7. It was subsequently shown that in periodic arrangements of ideally lossless dielectrics, the propagation of light can be totally suppressed at certain wavelengths

regardless of propagation direction and polarization.

翻译： 结果表明，在周期性排列的理想无损介质中，光的传播在一定波长下可以被完全抑制，而不受传播方向和偏振的影响。

8. The simplest device using the principles of photonic crystals is the one-dimensional photonic crystal, well known under the name of the Bragg mirror or the multilayer reflector.

翻译： 利用光子晶体原理最简单的器件是一维光子晶体，以布拉格反射镜或多层反射镜的名字而闻名。

9. With an appropriate choice of layer thickness and refractive indices, waves reflected from each interface are in phase, whereas waves transmitted are out of phase.

翻译： 在适当选择层厚和折射率的情况下，从每个界面反射的波是同相的，而传输的波是非相位的。

10. By solving the Schrodinger' s wave equation for a periodic potential, electron energy states separated by forbidden bands are obtained.

翻译： 通过求解薛定谔波动方程的周期性电势，可以获得被禁带分隔的电子能态。

11. When properly designed, the photonic crystal cladding running along the entire fiber length can prevent the escape of light from the hollow core.

翻译： 如果设计得当，可以沿着整个光纤长度的光子晶体包层防止光从空心芯逸出。

12. In fact, optical fibers would be the most suitable delivery means for many applications, but at present they are unusable, due to the fiber damage and the negative nonlinear effects caused by the high optical powers and energies, as well as to the fiber group-velocity dispersion, which disperses the short pulses.

翻译： 事实上，光纤将是许多应用中最合适的传输手段，但由于高功率和高能量导致的光纤损伤和负非线性效应，以及分散短脉冲的光纤群速度色散，使得光纤目前无法使用。

Unit 9
Applied Techniques

9.1 Optical Thin Film Technology

Optical thin films have been widely used in many different applications, controlling the reflection, transmission or adsorption of light as a function of wavelength. They can be grouped into two major categories based on the application. In the first, the light travels parallel to the plane of the substrate with the films acting as wave guides in the emerging field of integrated optics. Here light signals could replace electrical signals in applications such as communications and computers. In the second application, the light travels perpendicular to the film plane for use as antireflection coatings, edge filters, high efficiency mirrors, beam splitters, etc..

The term "thin" is used to indicate a layer whose thickness (perpendicular to the substrate) is the same order of magnitude as the wavelength of interest, and the extent (parallel to the substrate) is a very large number of wavelengths. Typical layers might range in thickness from 80 nm in the visible to twenty times that in the infrared. Filters are composed of a stack of such layers, alternating between high and low refractive indices, with typically 20~40 layers, although in some cases they may have one hundred layers or more. Thin film filters operate by interference of the light reflected from the various layers as the light passes through perpendicular to the substrate.

9.1.1 Design of Optical Thin Film

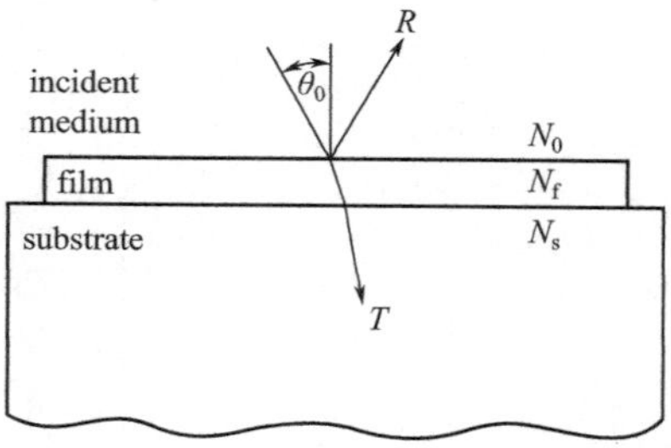

Figure 9-1 A single thin film

When light strikes a film, it can either be reflected, transmitted or lost to absorption or scattering. If we consider a light beam incident on a homogeneous parallel-sided film, the amplitude and polarization state of the light transmitted and reflected can be calculated in terms of the angle of incidence and the optical constants of the three materials involved, as shown in Figure 9-1. Light from the incident medium of refractive index N_0 passes through film material of index N, and enters substrate material of index N_s. The incident medium is often air with an index of refraction assumed to be equal to 1.0, the index of vacuum. The film and substrate materials can be transparent or absorbing in which the optical constant (or complex index of refraction), N, is given by

$$N = n - ik \tag{9.1}$$

where n is the refractive index and k is the extinction coefficient. At optical frequencies, $\varepsilon = n^2$, where ε is the dielectric constant of the material. The extinction coefficient is related to the absorption coefficient, α, by the expression

$$\alpha = 4\pi k/\lambda \tag{9.2}$$

where α determines the intensity I transmitted through an absorbing medium by the exponential law of absorption

$$I = I_0 e^{-\alpha x} \tag{9.3}$$

The reflectance and transmittance at the boundaries between these regions can be conveniently expressed in terms of the Fresnel coefficients. When the regions are absorbing, these terms are large and cumbersome but simplify into terms involving only n if the regions are transparent. Thus, if we assume that $k = 0$ and that plane waves strike a plane boundary at normal incidence, the reflectance between two regions is given by

$$R = \left(\frac{n_0 - n_s}{n_0 + n_s}\right)^2 \tag{9.4}$$

For example, the reflectance of uncoated window glass (with $n_s = 1.52$) in air (with $n_0 = 1.0$) is 4.3%. The transmittance, T, through the surface would be

$$T = 1 - R = 95.7\% \tag{9.5}$$

and through both surfaces would be 95.7% of 95.7% or 91.7%. By comparison, the transmittance through a single lens of $n = 1.9$ would be 81.7%. Complex systems may have many lenses cascaded together and the losses may total 50% or more. Infrared systems use very high index materials such as germanium which has a reflectance of 85% per surface. In addition to power lost, these reflections cause "ghost" images, thus it is necessary to put antireflection coatings on the lenses.

Antireflection coatings operate on the principle of interference of the light reflected from the front and back surfaces of the films. The optical thickness of a thin film is defined as the index n_f times the physical thickness d_f. In the case where $d_f = \lambda/4$, $3\lambda/4$, $5\lambda/4$, etc., the equation for calculating the reflectance of a single layer as in Figure 9-1 at normal incidence for the wavelength of interest λ_0 simplifies to

$$R = \frac{n_0 - \dfrac{n_f}{n_s}}{n_0 + \dfrac{n_f}{n_s}} \tag{9.6}$$

This is a very useful formula for estimating how well a particular single layer coating will do. We can calculate the film index needed for zero reflectance at one wavelength as

$$n_f = (n_0 n_s)^{1/2} \tag{9.7}$$

For a lens of $n = 1.9$, we get $n_f = 1.33$ which happens to be the index of refraction in the visible of MgF_2, the most widely used material for single layer antireflection coatings and one of the lowest index materials available. Thus, for glasses with an index lower than 1.9 (which is usually the case), we must either accept a little higher loss for the economy of a single layer MgF_2 coating or go to the additional cost of adding more layers of other materials. Adding more layers also gives the advantage of being able to achieve a lower reflectance over a much broader wavelength range.

The interference effect can be used by building up a stack of alternating high and low index materials to produce many interesting results. Figure 9-2 shows the construction of such a quarter wave stack in which the optical thickness $n_j d_j$ of each layer is again equal to one quarter the wavelength of interest. The multilayer is completely specified if we know n_j, k and d_j for each layer, n_0 for the incident medium plus n_s and k_s for the substrate. Given the angle of incidence θ_0, we can calculate the reflectance R and the transmittance T as a function of wavelength.

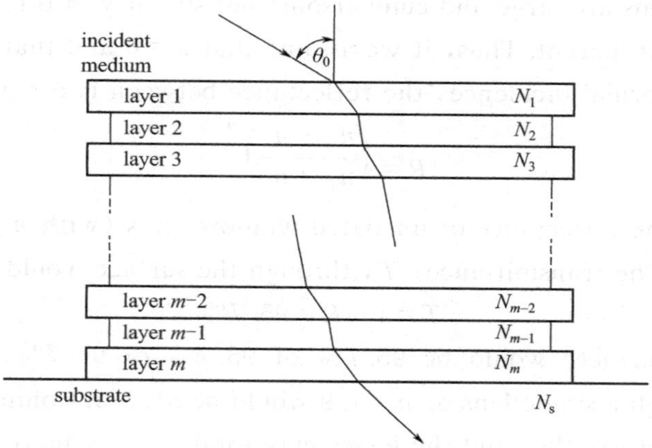

Figure 9-2 A multilayer stack

Figure 9-3 shows the reflectance vs. wavelength of six and twelve layer stacks of Si/SiO_2. They are plotted as a function of λ_0/λ, where the wavelength of interest $\lambda_0 = 4$. The multilayer has a characteristic stopband (or high-reflectance region) symmetric about the wavelength λ_0 surrounded by long and short-wave pass regions characterized by many ripples in the passbands. The width of the high reflectance region is determined by the ratio of the high to low index n_H/n_L. The higher the ratio, the wider the stopband. The maximum reflectance depends on the number of layers as well as the ratio n_H/n_L. The ripple in the passband is bounded by an envelope determined by n_H/n_L whereas the number of peaks

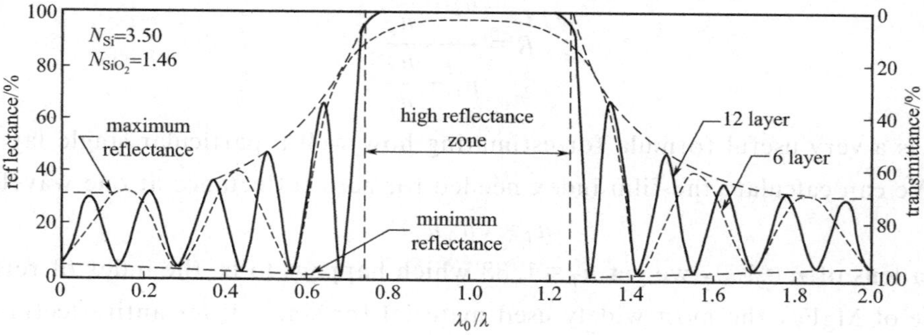

Figure 9-3 Reflectance vs. wavelength for Si/SiO_2 multilayer stacks

depends on the number of layers.

By suppressing the ripple on one side of the stopband we can create an edge fitter. A long-wave pass (LWP) filter transmits the long wavelengths and efficiently reflects the short ones. On the other hand, a short-wave pass (SWP) filter transmits the short wavelengths and reflects the long ones. A SWP filter is sometimes referred to as a high-pass filter meaning it passes high frequencies. Likewise, a LWP filter is sometimes called a low-pass filter and care must be used not to confuse these terms. The bandpass filter transmits a relatively narrow range of wavelengths and effectively rejects those on both sides.

Once a design has been made, the wavelength characteristics can be moved in wavelength by changing the layer thicknesses by the same ratio. For example, a LWP filter with an edge at 400 nm can be shied to 800 nm by making all the layers twice as thick.

As the angle of incidence increases, two effects are seen. First, the effective thickness of the layers becomes smaller which causes the filter to shift to shorter wavelength. Since the effect is inversely proportional to the film index, at large angles the layers are no longer matched at the same wavelength and the spectral shape degrades. The second effect at non-normal incidence is that the spectral characteristic becomes dependent on the polarization of the incident beam. This can be optimized and used to advantage for such components as polarizing beam splitters. But in general, these angle effects are detrimental, especially for a non-collimated beam.

9. 1. 2 Deposition of Optical Thin Film

There are many methods of thin film deposition. The major technique in use today for the deposition of optical films can be termed physical vapor deposition (PVD) in which the material condenses from the vapor phase onto the desired substrate. There are many different ways the vapor can be produced, but thermal evaporation (or vacuum evaporation) is the most widely used technique today. It is carried out in a vacuum to prevent unwanted chemical reactions and to keep the film from becoming porous by incorporating gas during the condensation process. The substrate to be coated is typically placed in a vacuum chamber, pumped down in the range of 1×10^{-6} torr (10^{-9} atmosphere) and heated to 200～300 ℃. The coating material is then heated until it evaporates or sublimes, travels across the vacuum chamber, then condenses on the substrate.

The condition and cleanliness of the substrate is critical to the quality of the coating. Since the forces which hold the film onto the substrate are all short range, even a few molecular layers of contaminant will result in a film with reduced adhesion. Since the layers are only a few tenths of a micron thick, they will faithfully reproduce any irregularities in the surface and cannot be expected to smooth over imperfections left from grinding and polishing. As a matter of fact, imperfections and especially stains in the glass surface that might be unobservable are greatly enhanced by coating. Once the substrates are cleaned and loaded into the fixtures they should be put under vacuum and coated as soon as possible.

The fixture is generally placed at the top of the chamber with the substrates held in by gravity. To improve thickness uniformity, the holder is often a dome which is rotated about the vertical axis of the chamber. By using uniformity masking, thickness from one area of the fixture to another can be controlled to ±2% or better. Tighter control may often be achieved with planetary tooling where smaller plates are rotated about their axes which are also moved about the chamber axis in a planetary motion. Some argue that the slight improvement in uniformity possible with the planetary is not worth the additional capital cost, complexity, particle generation and reduced throughput.

For vacuum evaporation, a high vacuum on the order of 10^{-6} torr is necessary so that evaporant molecules will travel across the chamber without encountering other molecules. This can be obtained with a variety of high vacuum pumps, the most common being the hot oil diffusion pump. It has been the workhorse of the industry for many years because it is economical, reliable and has a high throughput. If operated properly, it will not contaminate the chamber and substrates with oil. However, if oil contamination is seen, it is most likely from the mechanical pump used to "rough down" the chamber and back up the high vacuum pump. In recent years, cryopumps and turbomolecular pumps have sometimes been used in place of diffusion pumps because of the perception that they will eliminate oil contamination.

Rough pumping takes on the order of 5～10 minutes with high vacuum pumping taking an hour or more. During this time the substrates are usually heated to improve adhesion and film properties. The heaters are either the rod type placed behind the substrates or, more often, quartz lamps placed below the substrates. In both cases, heating is by radiation in the vacuum.

The two most widely used methods of evaporant heating are the resistance source and electron beam gun. In resistance heating, a large current of a few hundred amperes at a low voltage perhaps passes through a boat or filament to evaporate or sublime the material. The electron beam gun generates a stream of electrons of 6～10 kV up to a few amperes. Using magnetic fields, the beam is deflected typically 270 ℃ to keep the filament out of the path of the evaporant which greatly extends the filament lie. Electromagnets are used to move the beam around on the source material to control what area gets heated, as well as to raster the beam to avoid hot spots and keep the surface of the evaporant level. If this is not done, the surface of the evaporant changes shape and the distribution of deposited film changes.

As implied earlier, the optical thickness (index times physical thickness) must be controlled accurately for each layer. There have been many schemes devised for controlling thickness, but the optical monitor and the quartz crystal microbalance are the two methods which are most widely used today.

The optical monitor measures the change in reflected or transmitted light from a substrate or glass witness slide located near the work, usually in the center of the chamber. As the thickness increases, the signal goes through maxima and minima corresponding to multiples of quarterwaves. Since the optical monitor measures optical thickness, it compensates for slight changes in index of refraction.

The quartz crystal monitor measures the change in mass of a vibrating quartz crystal as the coating builds up on it. The advantage is that the output signal varies linearly with thickness and thus it is easy to build this into a controller for both deposition rate and final thickness. The quartz crystal controller is excellent for metal films, but there is some debate over which method is best for dielectric optical thin films. Many have chosen to use the quartz crystal monitor to control the deposition cycle including rate, but use an optical monitor to control final thickness, especially for filters requiring high precision.

9.2 Photolithography

Photolithography, or optical lithography is a process used in microfabrication to selectively remove parts of a thin film or the bulk of a substrate. It uses light to transfer a geometric pattern from a photo mask to a light-sensitive chemical photoresist, or simply resist, on the substrate.

Optical lithography is a fascinating field. It requires knowledge of geometrical and wave optics, optical and mechanical systems, diffraction imaging, Fourier optics, resist systems and processing, quantification of imaging performance, and the control of imaging performance. Even the history of its development helps to stimulate new ideas and weed out less promising ones. Practitioners of optical lithography may only have a vague idea of its theory, and likewise, theoreticians may not have the opportunity to practice the technology on the manufacturing floor. In this section, we will briefly introduce the basic procedure and predict the new trend of photolithography.

Optical lithography has been the workhorse of semiconductor fabrication since the inception of integrated circuits. The lensless proximity printing system gradually gave way to projection-printing systems, and one-to-one replication systems became reduction systems. It took this latest form from the 0.15NA 436 nm g-line lens, featuring resolution over 2 μm with a 0.8k_1 factor, all of the way through raising the NA until the lens became too expensive to build at that time, reducing wavelength to reposition the NA for the next round of increases, and lowering k_1 whenever the pace of NA and wavelength changes are behind the circuit shrinking roadmap, as shown in Figure 9-4.

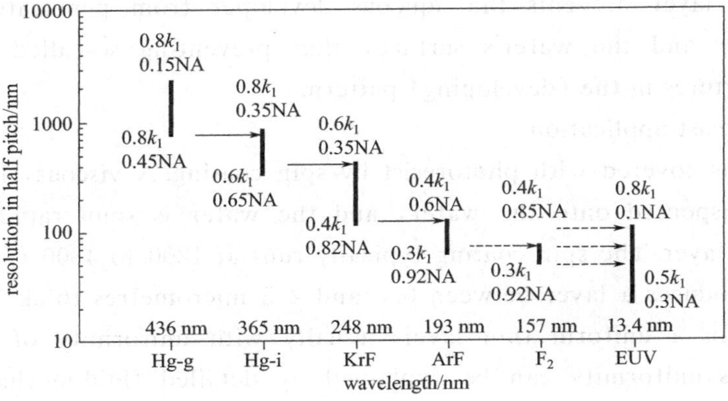

Figure 9-4 Evolution of projection optical lithography from a 0.15NA g-line

9.2.1 Basic Procedure

A single iteration of photolithography combines several steps in sequence, as shown in Figure 9-5. Modern cleanrooms use automated, robotic wafer track systems to coordinate the process. The procedure described here omits some advanced treatments, such as thinning agents or edge-bead removal.

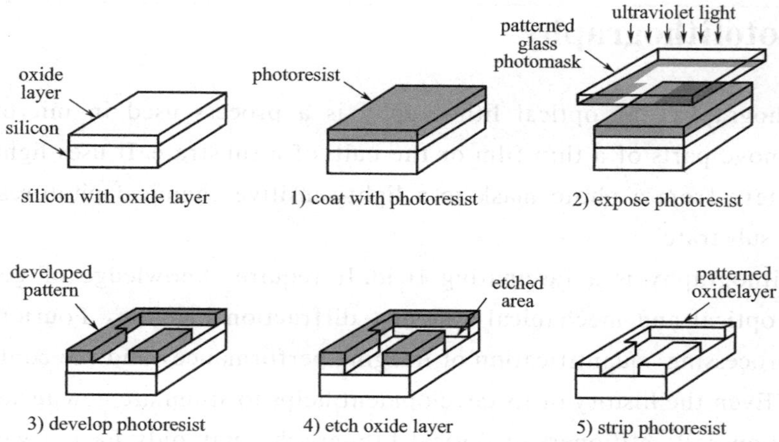

Figure 9-5 Basic photolithography process

(1) Cleaning

If organic or inorganic contaminations are present on the wafer surface, they are usually removed by wet chemical treatment.

(2) Preparation

The wafer is initially heated to a temperature sufficient to drive off any moisture that may be present on the wafer surface. Wafers that have been in storage must be chemically cleaned to remove contamination. A liquid or gaseous "adhesion promoter" is applied to promote adhesion of the photoresist to the wafer. The surface layer of silicon dioxide on the wafer reacts with HMDS (Hexamethyl Disilazane) to form tri-methylated silicon-dioxide, a highly water repellent layer not unlike the layer of wax on a car's paint. This water repellent layer prevents the aqueous developer from penetrating between the photoresist layer and the wafer's surface, thus preventing so-called lifting of small photoresist structures in the (developing) pattern.

(3) Photoresist application

The wafer is covered with photoresist by spin coating. A viscous, liquid solution of photoresist is dispensed onto the wafer, and the wafer is spun rapidly to produce a uniformly thick layer. The spin coating typically runs at 1200 to 4800 r/min for 30 to 60 seconds, and produces a layer between 0.5 and 2.5 micrometres thick. The spin coating process results in a uniform thin layer, usually with uniformity of within 5 to 10 nanometres. This uniformity can be explained by detailed fluid-mechanical modeling, which shows that the resist moves much faster at the top of the layer than at the bottom, where viscous forces bind the resist to the wafer surface. Thus, the top layer of resist is quickly ejected from the wafer's edge while the bottom layer still creeps slowly radially along the wafer. In this way, any 'bump' or 'ridge' of resist is removed, leaving a very flat layer. Final thickness is also determined by the evaporation of liquid solvents from the resist. For very small, dense features ($<$125 nm or so), thinner resist thicknesses ($<$0.5 micrometres) are needed to overcome collapse effects at high aspect ratios; typical aspect ratios are $<$4 : 1.

The photoresist-coated wafer is then prebaked to drive off excess photoresist solvent, typically at 90 to 100 ℃ for 30 to 60 seconds on a hotplate.

(4) Exposure and developing

After prebaking, the photoresist is exposed to a pattern of intense light. Optical lithography typically uses ultraviolet light. Positive photoresist, the most common type, becomes soluble in the basic developer when exposed; exposed negative photoresist becomes insoluble in the (organic) developer. This chemical change allows some of the photoresist to be removed by a special solution, called "developer" by analogy with photographic developer.

(5) Etching

In etching, a liquid ("wet") or plasma ("dry") chemical agent removes the uppermost layer of the substrate in the areas that are not protected by photoresist. In semiconductor fabrication, dry etching techniques are generally used, as they can be made anisotropic, in order to avoid significant undercutting of the photoresist pattern. This is essential when the width of the features to be defined is similar to or less than the thickness of the material being etched (i. e. when the aspect ratio approaches unity). Wet etch processes are generally isotropic in nature, which is often indispensable for, microelectromechanical systems, where suspended structures must be "released" from the underlying layer. The development of low-defectivity anisotropic dry-etch process has enabled the ever-smaller features defined photolithographically in the resist to be transferred to the substrate material.

(6) Photoresist removal

After a photoresist is no longer needed, it must be removed from the substrate. This usually requires a liquid "resist stripper", which chemically alters the resist so that it no longer adheres to the substrate. Alternatively, photoresist may be removed by a plasma containing oxygen, which oxidizes it. This process is called ashing, and resembles dry etching.

9.2.2 New Trend of Photolithography

Optical lithography is marching toward the end of its legendary longevity in being a semiconductor-manufacturing workhorse. After using the 193 nm wavelength and wafer immersion, polarized illumination, and improved masks, the only way to stretch optical lithography farther is to split the pitch by double patterning to reduce k_1 to 0.15 from 0.3, suggesting a half pitch of 21.5 nm. However, starting from a 32 nm half pitch, there are opportunities for other technologies to compete with 193 nm immersion.

We now compare the challenges of optical lithography as it faces its end and its possible contenders, which consist of EUV, MEB ML2 systems, and nanoimprint.

(1) EVU lithography

EUV lithography (EUVL) promises single-exposure, simple optical proximity correction (OPC) and has commanded worldwide development efforts at an unprecedented running cost in addition to hundreds of million dollars of sunk cost.

Will it be cost effective? The system is inevitably huge due to the need for high-speed reticule and wafer stages in vacuum. No doubt, the production-worthy hardware will be extremely expensive. The key to suppression of wafer exposure cost is to improve throughput. With resist sensitivity staying at a reasonable level of the order of 30 mJ/cm^2 to reduce line-edge roughness and shot-noise effects, the only avenue for high throughput is a high-source power, which also tends to be expensive and environmentally unfriendly, if it can even be made to work.

EUV resists may not provide cost relief. Even if the material production costs were contained, the EUV exposure tools that resist companies will need to purchase inevitably adds to the cost of EUV resists. The high cost of a defect-free EUV mask substrate is well known. The cost of a mask inspection may become a significant factor because actinic light is needed. The cost of yield loss and efforts to prevent it, due to exposure without a pellicle and radiation-induced mask contamination, has yet to be determined from field exposures at manufacture-worthy throughput.

In short, even if every technical problem in EUV lithography was solved, the high exposure cost and high raw-energy consumption may prove detrimental.

(2) MEB ML2

E-beam maskless lithography also has decades of history. It has not been the industry's workhorse for wafer imaging, except for making personalized interconnects in the earlier days at IBM. With circuit complexity ever increasing, the improvement of throughput, even though impressive by itself, could not catch up. E-beam direct write did become a workhorse for mask making.

To increase the e-beam throughput to a cost-effective level, a high degree of parallelism must be used. It can be in the number of minicolumns, the number of MEMS-fabricated microlenses, pixels in a programmable mask, or numbers of MEMS-fabricated apertures. The exploration of parallelism need not stop here. Some of the above can be favorably combined. Wafers can be exposed simultaneously in a cluster. Without clustering, throughputs on the order of 10 to 40 wph have been claimed. Some are dependent on pattern density due to space-charge-limited current. Some are not.

Even though e-beams have extremely high resolution and depth of focus, scattering in a resist spreads a well-focused beam and fundamentally limits resolution. The only way to reduce scattering is to use a smaller resist thickness. It is foreseeable that resist thickness on the order of 50 nm or less must be used. A multilayer resist system or highly selective hard mask must be used.

The data rate of MEB ML2 is quite demanding. Heavy-duty data processing equipment is needed. Much of the cost in optics for an optical scanner is transferred to data-processing equipment. Fortunately, the cost reduction of data-processing equipment predictably follows Moore's law.

(3) Nanoimprint lithography

Nanoimprint, the lensless replication technique, seems to have unlimited resolution capability and the molding tool costs mere pennies. Nanoimprint has taken contact printing to a higher level. Instead of letting the mask contact the photoresist while still occupying its own space, the imaging medium is mingled with the molding template, i. e., the mask in nanoimprint. Just like Contac printing, nanoimprint is susceptible to defects. Mingling the medium with the mold does not offer promise of fewer defects.

The mold has a much shorter lifespan than the optical mask, but child molds can be replicated effortlessly using the same molding technique. Grandchildren can also be produced. Children and grandchildren must be stored and managed.

(4) Comparison of the three technologies

EUVL systems have the largest momentum and highest expectation to succeed. However, the list of challenges is still long. The higher the development cost, the more financial burden the industry will must bear, regardless of its success or failure.

MEB ML2 systems have the potential to be cost effective. They also have a long list of challenges; however, it is worth a try based on the relatively low cost to develop, the potential of becoming a workhorse, and for the absence of an increasingly difficult mask infrastructure.

Nanoimprint holds the promise of high resolution and a high height-to-width aspect ratio. Defects, template fabrication, template lifetime, throughput, and cost remain tough problems to solve.

It is not inconceivable that ArF water-immersion will be holding the last frontier for lithography, especially if double-mask exposure and high-contrast resists are realized. There may be more lithography-friendly designs, better semiconductor devices, and circuit optimization for the last frontier.

9.3 Biophotonics

Photonics is an all-encompassing light-based optical technology that is being hailed as the dominant technology for this new millennium. The invention of lasers, a concentrated source of monochromatic and highly directed light, has revolutionized photonics. Since the demonstration of the first laser in 1960, laser light has touched all aspects of our lives, from home entertainment, to high-capacity information storage, to fiber-optic telecommunications, thus opening up numerous opportunities for photonics.

A new extension of photonics is biophotonics, which involves a fusion of photonics and biology. Biophotonics deals with interaction between light and biological matter. A general introduction to biophotonics is illustrated in Figure 9-6.

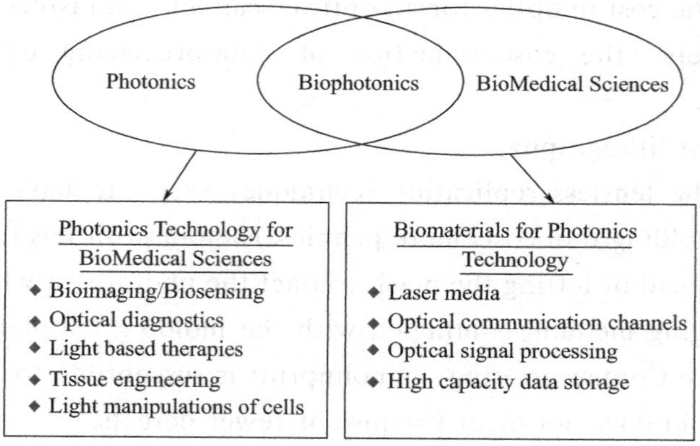

Figure 9-6 Biophotonics as defined by the fusion of photonics and biomedical sciences

The two broad aspects of biophotonics are also identified.

The use of photonics for optical diagnostics, as well as for light-activated and light-guided therapy, will have a major impact on health care. This is not surprising since nature has used biophotonics as a basic principle of life from the beginning. Harnessing photons to achieve photosynthesis and conversion of photons through a series of complex steps to create vision are the best examples of biophotonics at work. Conversely, biology is also advancing photonics, since biomaterials are showing promise as new photonic media for technological applications.

As an increasingly aging world population presents unique health problems, biophotonics offers great hope for the early detection of diseases and for new modalities of light-guided and light-activated therapies. Lasers have already made a significant impact on general, plastic, and cosmetic surgeries. Two popular examples of cosmetic surgeries utilizing lasers are skin resurfacing (most commonly known as wrinkle removal) and hair removal. Laser technology also allows one to administer a burst of ultrashort laser pulses that have shown promise for use in tissue engineering. Furthermore, biophotonics may produce retinal implants for restoring vision by reverse engineering nature's methods.

An overview of biophotonics for health care applications is presented in Figure 9-7. It illustrates the scope of biophotonics through multidisciplinary comprehensive research and development possibilities.

9.3.1 Bioimaging

Biomedical imaging has become one of the most relied-upon tools in health care for diagnosis and treatment of human diseases. The evolution of medical imaging from plain radiography (radioisotope imaging), to X-ray imaging, to computer-assisted tomography (CAT) scans, to ultrasound imaging, and to magnetic resonance imaging (MRI) has led to revolutionary improvements in the quality of healthcare available today to our society. However,

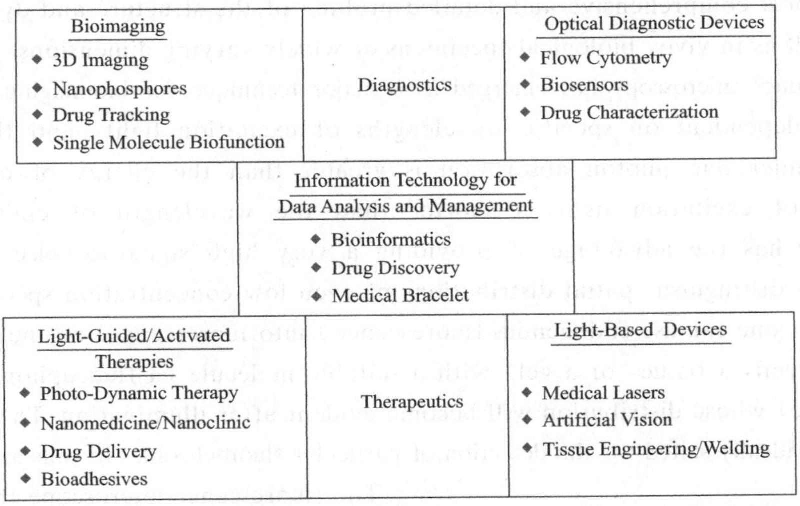

Figure 9-7 The comprehensive multidisciplinary scope of biophotonics for health care

these techniques are largely focused on structural and anatomical imaging at the tissue or organ levels. In order to develop novel imaging techniques for early detection, screening, diagnosis, and image-guided treatment of life-threatening diseases and cancer, there is a clear need for extending imaging to the cellular and molecular biology levels. Only information at the molecular and cellular levels can lead to the detection of the early stages of the formation of a disease or cancer or early molecular changes during intervention or therapy.

The currently used medical techniques of X-ray imaging, radiography, CAT scans, ultrasound imaging, and MRI have a number of limitations.

Optical imaging overcomes many of these deficiencies. Contrary to the perception based on the apparent opacity of skin, light, particularly in near-IR region, penetrates deep into the tissues. Furthermore, by using a minimally invasive endoscope fiber delivery system, one can reach many organs and tissue sites for optical imaging. Thus, one can even think of an "optical body scanner" that a physician may use some day for early detection of a cancer or an infectious disease.

Optical imaging utilizes the spatial variation in the optical properties of a biospecies, whether a cell, a tissue, an organ, or an entire live object. The optical properties can be reflection, scattering, absorption, and fluorescence. Therefore, one can monitor spatial variation of transmission, reflection, or fluorescence to obtain an optical image. The use of lasers as an intense and convenient light source to generate an optical response, whether reflection, transmission, or emission, has considerably expanded the boundaries of optical imaging, making it a most powerful technique for basic studies as well as for clinical diagnostics.

The area of optical imaging is very rich, both in terms of the number of modalities and with regard to the range of its applications. It is also an area of very intense research worldwide because new methods of optical imaging, new, improved, and miniaturized instrumentations, and new applications are constantly emerging.

Fluorescence microscopy is the most widely used technique for optical bioimaging. It provides a most comprehensive and detailed probing of the structure and dynamics for in vitro, as well as in vivo, biological specimens of widely varying dimensions.

Fluorescence microscopy has emerged as a major technique for bioimaging. Fluorescence emission is dependent on specific wavelengths of excitation light, and the energy of excitation under one photon absorption is greater than the energy of emission (the wavelength of excitation light is shorter than the wavelength of emission light). Fluorescence has the advantage of providing a very high signal-to-noise ratio, which enables us to distinguish spatial distributions of even low concentration species. To utilize fluorescence, one can use endogenous fluorescence (autofluorescence) or one may label the specimen (a cell, a tissue, or a gel) with a suitable molecule (a fluorophore, also called fluorochrome) whose distribution will become evident after illumination. The fluorescence microscope is ideally suited for the detection of particular fluorochromes in cells and tissues.

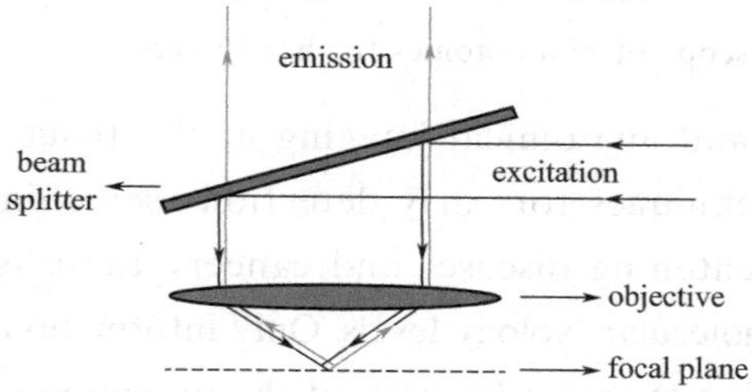

Figure 9-8 Basic principle of epi-fluorescence illumination

The fluorescence microscope that is in wide use today follows the basic "epifluorescence excitation" design utilizing filters and a dichroic beam splitter. The object is illuminated with fluorescence excitation light through the same objective lens that collects the fluorescence signal for imaging. A beam splitter, which transmits or reflects light depending on its wavelength, is used to separate the excitation light from the fluorescence light. In the arrangement, shown in Figure 9-8, the shorter-wavelength excitation light is reflected while the longer-wavelength emitted light is transmitted by the splitter.

With the advent of different fluorochromes/ fluorophores, specifically targeting different parts of the cells or probing different ion channel processes (e. g., Ca^{2+} indicators), the fluorescence microscopy has had a major impact in biology. The development of confocal microscopy has significantly expanded the scope of fluorescence microscopy.

9. 3. 2 Optical Biosensors

The field of biosensors has emerged as a topic of great interest because of the great need in medical diagnostics and, more recently, the worldwide concern of the threat of chemical and bioterrorism. The constant health danger posed by new strands of microbial organisms and spread of infectious diseases is another concern requiring biosensing for detecting and identifying them rapidly. Optical biosensors utilize optical techniques to detect and identify chemical or biological species. They offer a number of advantages such as the ability for principally remote sensing with high selectivity and specificity and the ability to use unique biorecognition schemes.

Biosensors are analytical devices that can detect chemical or biological species or a microorganism. They can be used to monitor the changes in vivo concentrations of an endogenous specie as a function of a physiological change induced internally or by invasion of a microbe. Of even more recent interest is the use of biosensors to detect toxins, bacteria, and viruses because of the danger posed by chemical and biological terrorism. Biosensors thus find a wide range of applications, such as clinical diagnostics, drug development, environmental monitoring (air, water, and soil), and food quality control, etc..

A biosensor in general utilizes a biological recognition element that senses the presence of an analyte (the specie to be detected) and creates a physical or chemical response that is converted by a transducer to a signal. The general function of a biosensor system is described in Figure 9-9. The sampling unit introduces an analyte into the detector and can be as simple as a circulator. The recognition element binds or reacts with a specific analyte, providing biodetection specificity. Enzymes, antibodies or even cells such as yeast or bacteria have been used as biorecognition elements. Stimulation, in general, can be provided by optical, electric, or other kinds of force fields that extract a response as a result of biorecognition. The transduction process transforms the physical or chemical response of biorecognition, in the presence of an external stimulation, into an optical or electrical signal that is then detected by the detection unit. The detection unit may include pattern recognition for identification of the analyte. In the most commonly used form of an optical biosensor, the stimulation is in the form of an optical input. The transduction process induces a change in the phase, amplitude, polarization, or frequency of the input light in response to the physical or chemical change produced by the biorecognition process. Some of the other approaches use electrical stimulation to produce optical transduction (e. g., an electroluminescent sensor) or an optical stimulation to produce electrical transduction (e. g., a photovoltaic sensor).

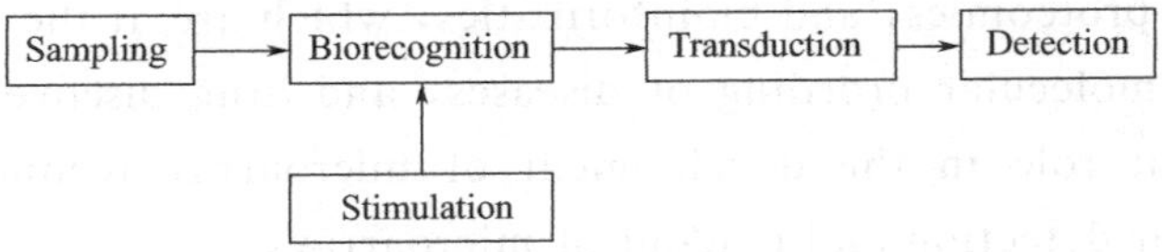

Figure 9-9　General scheme for biosensing

The field of biosensors has been active over many decades. The earlier successes were sensors utilizing electrochemical response (Janata, 1989). This type of sensor still tends to dominate the current commercial market. However, progress in fiber optics and integrated optics (such as channel waveguides and surface plasmon waves) and the availability of microlasers (solid-state diode lasers) have made optical biosensors a very attractive alternative for many applications. An optical biosensor, in general, utilizes a change in the amplitude (intensity), phase, frequency or polarization of light created by a recognition element in response to a physiological change or the presence of a chemical or a biologic (e. g., microorganism). Enhancement of the sensitivity and selectivity of the optical response is achieved by immobilizing the biorecognition element (such as an antibody or an enzyme) on an optical element such as a fiber, a channel waveguide, or a surface plasmon propagation where light confinement produces a strong internal field or an evanescent (exponentially decaying) external field. Thus, the main components of an optical biosensor include ① a light source, ② an optical transmission medium (fiber, waveguide, etc.), ③ immobilized biological recognition element (enzymes, antibodies or microbes), ④ optical probes (such as a fluorescent marker) for transduction, and ⑤ an optical detection system.

Wadkins et al. (1998) used a scheme, shown in Figure 9-10, that used patterned antibody channels.

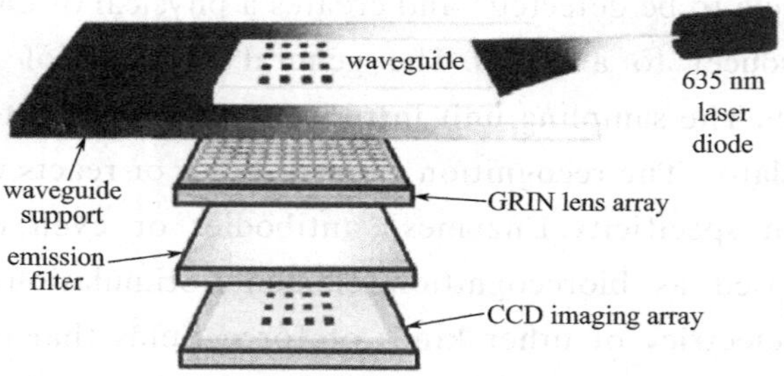

Figure 9-10 Array biosensors developed by Ligler, Golden, and co-workers at the Naval Research Laboratory

They demonstrated the detection of Y. pestis F1 (a Gram-negative rod-shaped bacterium) in clinical fluids such as whole blood, plasma, urine, saliva, and nasal secretion.

9. 3. 3 Microarray Technology for Genomics and Proteomics

Microarray technology provides a powerful tool for high-throughput rapid analysis of a large number of samples. This capability has been of significant value in advancing the fields of genomics, proteomics, and bioinformatics, which are at the forefront of modern structural biology, molecular profiling of diseases, and drug discovery. Biophotonics has played an important role in the development of microarray technology, since optical methods are used for detection and readout of microarrays.

Microarray technology is a powerful, universally applicable analytical tool for rapid and simultaneous detection and analysis of a large number of biological assemblies with high sensitivity and precision. Thousands of DNA and RNA species can be simultaneously probed using a DNA microarray to provide detailed insight into cellular phenotyping and genotyping. Phenotyping is the process of identifying cells with specific markers. When antibodies are used, the process is called immunophenotyping. When genetic constructs are used, it is called genotyping.

The various types of biological microarrays are compiled in Figure 9-11. Of these the DNA microarray technology is most developed and in wide usage all around the world. The term gene chip or biochip is also often used to describe a DNA microarray because the approaches used to fabricate DNA microarrays often involve processing analogous to that used to produce semiconductor microchips. Since its commercial availability in 1996, DNA microarrays have now become a major tool for genomics and drug discovery.

Protein microarrays or protein chips utilize microarrays of immobilized fusion proteins (proteins that fuse with other proteins) or antibodies. However, the protein chips currently are not sufficiently robust for high throughput studies.

Cell microarrays are a relatively new development that utilizes live cells expressing a c-DNA of interest.

Tissue microarrays developed in the laboratory of Kallioniemi provides a new high-throughput tool for the study of gene dosage and protein expression patterns in a large number of individual tissues. This tool can provide a rapid and comprehensive molecular profiling of cancer and other diseases without exhausting limited tissue resources.

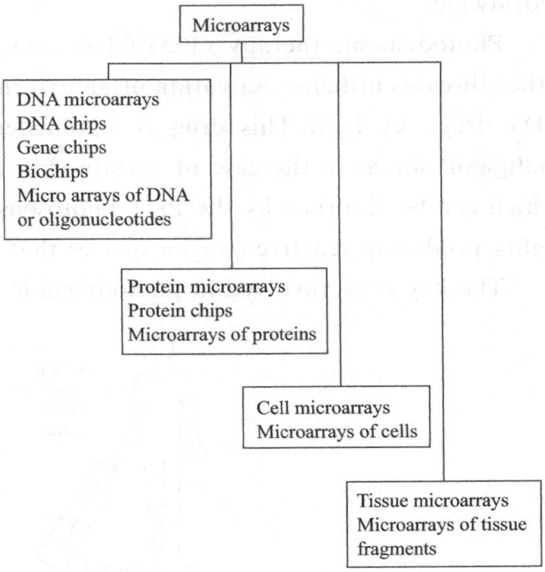

Figure 9-11 Various biological microarrays

The two main pieces of hardware for microarray technology are ① the microarray slide spotter and ② the microarray scanner. The two main approaches used to fabricate DNA microarrays are described in Figure 9-12.

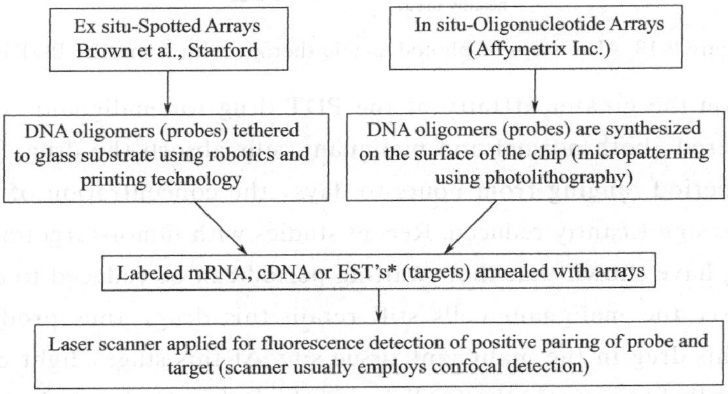

Figure 9-12 Two main approaches for DNA microarray technology

9.3.4 Light-Activated Therapy: Photodynamic Therapy

An important area of biophotonics is the use of light for therapy and treatment. The use of light to activate a photosensitizer eventually leads to the destruction of cancer or a diseased cell. This procedure is called photodynamic therapy (abbreviated as PDT) and constitutes a multidisciplinary area that has witnessed considerable growth in activities worldwide.

Photodynamic therapy (PDT) has emerged as a promising treatment of cancer and other diseases utilizing activation of an external chemical agent, called a photosensitizer or PDT drug, by light. This drug is administered either intravenously or topically to the malignant site as in the case of certain skin cancers. Then light of a specific wavelength, which can be absorbed by the PDT photosensitizer, is applied. The PDT drug absorbs this light, producing reactive oxygen species that can destroy the tumor.

The key steps involved in photodynamic therapy are shown in Figure 9-13.

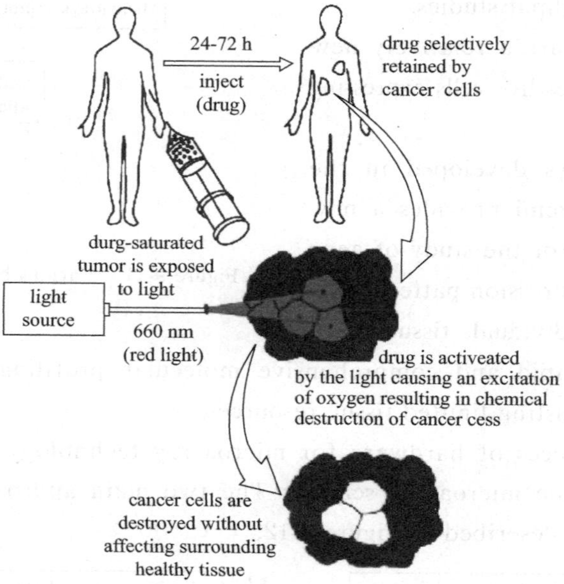

Figure 9-13 The steps of photodynamic therapy with a specific PDT drug

PDT relies on the greater affinity of the PDT drug for malignant cells. When a PDT drug is administered, both normal and malignant cells absorb the drug. However, after a certain waiting period ranging from hours to days, the concentration of the PDT drug in the normal cell is significantly reduced. Recent studies with tumor-targeting agents attached to the PDT drug have shown that their waiting period can be reduced to a matter of a few hours. In contrast, the malignant cells still retain this drug, thus producing a selective localization of this drug in the malignant tissue site. At this stage, light of an appropriate wavelength is applied to activate the PDT drug, which then leads to selective destruction of the malignant tissue by a photochemical mechanism (nonthermal, thus no significant local heating). In the case of cancer in an internal organ such as a lung, light is administered using a minimally invasive approach involving a flexible fiberoptic delivery system. In the case of a superficial skin cancer, a direct illumination method can endoscopic be used. Since coherence property of light is not required, any light source such as a lamp or a laser beam can be used. However, to achieve the desired power density at the required wavelength, a laser beam is often used as a convenient source for this treatment. The use of a laser beam also facilitates fiber-optic delivery.

There are only two currently approved drugs: photofrin and verteroporfin. In all cases, photofrin has been used as a photosensitizer that is activated at 630 nm. Fluorescence images show a tumor on the back before and after irradiation with PDT light. Drug dose: 5 mg/kg b. w. ; uptake time: 12 h; excitation light: 660 nm. This PDT drug is administered by an intravenous.

9. 3. 5 Nanotechnology for Biophotonics: Bionanophotonics

Some describe us as living in an era of Nanomania where there is a general euphoria about nanoscale science and technology. The fusion of nanoscience and nanotechnology with biomedical research has also broadly impacted biotechnology. Imagine nanosubmarines navigating through our bloodstreams and destroying nasty viruses and bacteria. Imagine nanorobots hunting for cancer cells throughout our body, finding them, then reprogramming or destroying them. A subject of science fiction at one time has now been transformed into a future vision showing promise to materialize. The fusion of nanoscience and nanotechnology into biomedical research has brought in a true revolution that is broadly impacting biotechnology. New terms such as nanobioscience, nanobiotechnology, and nanomedicine have come into existence and gained wide acceptance.

Nanophotonics is an emerging field that deals with optical interactions on a scale much smaller than the wavelength of light used (Shen et al., 2000). The three major areas of nanophotonics are shown in Figure 9-14.

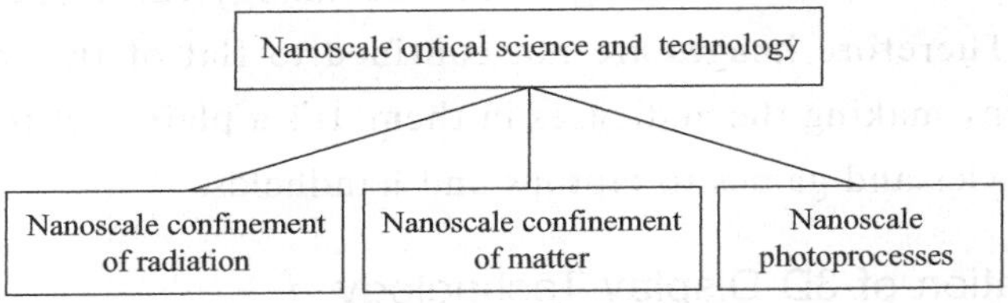

Figure 9-14 Three major areas of nanophotonics

Nanoscale confinement of radiation is achieved in a near-field geometry. This allows one to break diffraction barriers and obtain optical resolution to less than 100 nm. Near-field microscopy is becoming a powerful biomedical research tool to probe structure and functions of submicron dimension biological species such as bacteria. Nanoscale confinement of matter is achieved by producing nanoparticles, nanomers, nanodomains, and nanocomposites. The nanosize manipulation of molecular architecture and morphology provides a powerful approach to control the electronic and optical properties of a material. An example is a semiconductor quantum dot, a nanoparticle whose electronic band gap and thus the emission wavelength are strongly dependent on its size. Nanoscale control of the local structure in a nanocomposite, consisting of many domains and separated only on the nanometer scale, provides an opportunity to manipulate excitedstate dynamics and electronic energy transfer from one domain to another. Such nanostructured materials can provide significant benefits in fluorescence resonance energy transfer (FRET) imaging and in flow cytometry respectively.

Nanoscale photoprocesses such as photopolymerization provide opportunities for nanoscale photofabrication. Near-field lithography can be used to produce nanoarrays for DNA or protein detection. The advantage over the microarray technology is the higher density of arrays obtainable using near-field lithography, thus allowing one to use small quantities of samples. This is a tremendous benefit for protein analysis in the case when the amount of protein produced is very minute and there is no equivalent of DNA PCR amplification for proteins to enhance the detection.

The applications of nanophotonics to biomedical research and biotechnology range from biosensing, to optical diagnostics, to light activated therapy. Nanoparticles provide a highly useful platform for intracellular optical diagnostics and targeted therapy. The area of usage of nanoparticles for drug delivery has seen considerable growth.

Levy has developed the concept of a nanoclinic, a complex surface functionalized silica nanoshell containing various probes for diagnostics and drugs for targeted delivery.

Nanoclinics provide a new dimension to targeted diagnostics and therapy. These nanoclinics are produced by multistep nanochemistry in a reverse micelle nanoreactor.

9.4 3D Display Technology

In "3D", "D" is the first letter of "dimension", and "3D" means three-dimensional space. Comparing with ordinary 2D images, 3D technology can make the picture become stereo and lifelike. Therefore images are not confined to flat of the screen, as if walking outside of the screen, making the audiences in them. It's a plain fact that 3D is everywhere these days from movies and games to laptops and handhelds.

9.4.1 Classification of 3D Display Technology

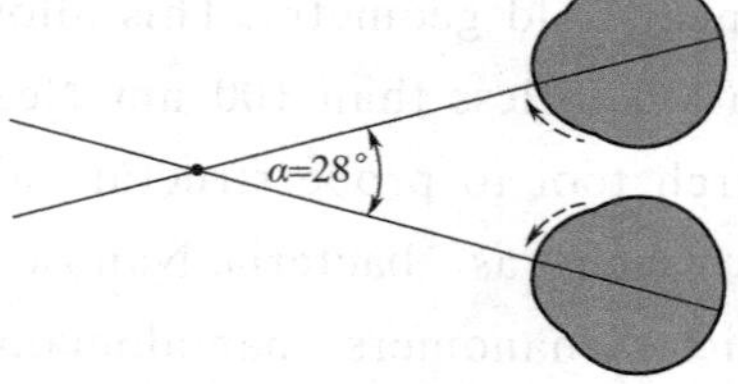

Figure 9-15 Interocular distance

Close just your right eye. Now just your left. Now your right. It's like in Wayne's World: camera one, camera two. You must have noticed that things change position a bit. This, of course, is because your eyes are a few inches apart; this is called the "interocular distance" and it varies from person to person. Note also that when you look at something close, objects appear in double in the background. Look at the corner of the screen. You can see the chair and window back there are doubled because you're actually rotating your eyes so they both point directly at what you're focusing on. This is called "convergence," as Figure 9-15 shows, and it creates a sort of X, the center of the X being what's being focused on.

So the objects at the center of the X are aligned at the same points on your respective retinas, but because of the interocular, that means that things in front and behind of that X are going to hit different points on those retinas, resulting in a double image. You can see through the double image because what's blocked for one eye is seen by the other, though from a slightly different point of view.

Next we have focus. Your eye focuses differently from most cameras, but the end result is similar. Try holding your finger out at arm's length and focusing on it, then on something distant that is directly behind it. Obviously the focus changes and you may have noticed that while the distant object was blurry while you looked at your finger, your finger (in double) was in focus while you looked at the distant object. That has to do with the optic qualities of your eye, which is important here. More important, however, is this: try it again, and this time pays attention to the feeling in your eye. You can feel that sort of like something's moving but you can't tell exactly what. Your eyes are rotating a bit, but only a few degrees, in order to converge further out, but more importantly, you're feeling the muscles of your eye actually crush or stretch out the lens of the eye, changing the path and internal focal plane of the light entering your eye. Farsightedness and nearsightedness occur at this stage, when either the lens or the eyeball itself is misshaped, resulting in focus being skewed or difficult to resolve one way or another.

Once the two images have been presented to your retinas, they pass back through the optic nerve to various visual systems, where an incredibly robust real-time analysis of the raw data is performed by several areas of the brain at once. Some areas look for straight lines, some for motion, some perform shortcut operations based on experience and tell you that yes indeed, the person did go behind that wall, they did not disappear into the wall, and that sort of thing. Eventually (within perhaps 50 milliseconds) all this information filters up into your consciousness and you are aware of color, depth, movement, patterns, and distinct objects within your field of view, informed mainly by the differences between the images hitting each of your retinas. It's important to note that vision is a learned process, and these areas in your visual cortex are "programmed" by experience as much as by anatomy and, for lack of a better term, instinct.

Although 3D display technologies are various, the most basic image-forming principle is similar. Figure 9-16 shows that these technologies use the human left and right eyes to receive different frames respectively, then with superposing and reconstructing image information, the brain form a fore-back, up-down, left-right, far-near image which is three dimensional.

Free viewing 3D display technology is now used mainly at public business occasions, and it will be applied to mobile phones and other portable devices. While in the field of family consumption, whether displays, projector or TV, it needs to cooperate with 3D glasses.

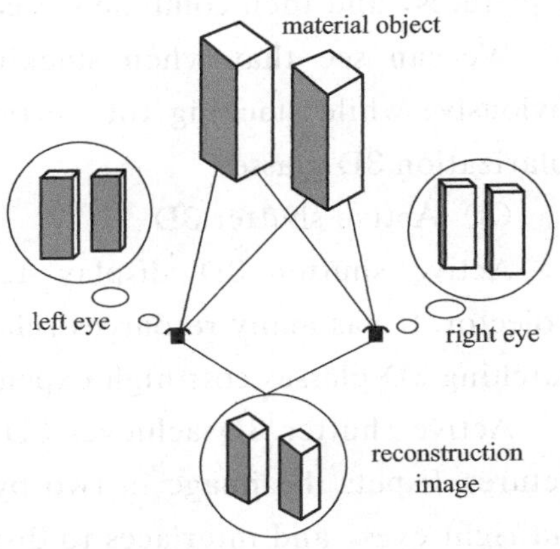

Figure 9-16 The basic image-forming principle demonstration

9.4.2 Aided-viewing

The aided-viewing 3D display technology includes three main types: anaglyphic 3D, polarization 3D and active shutter 3D. They are also usually called the method of colour division, light division and time division.

(1) Anaglyphic 3D

Anaglyphic 3D display technology need to use red-blue (or red-green) color 3D filter glasses to aid. The history of this technology is the longest, and the principle of image-forming is the simplest. It only takes a few dollars, while its 3D display effect is the worst.

Firstly, anaglyphic 3D gives light spectrum by the rotating filter wheel. Then, it uses optical filter with different color to filter the light, making a picture produce two images. Thus the person's each eye can see different images. However, this method will produce partial color picture edge.

(2) Polarization 3D

Polarization 3D is used together with the passive polarized glasses, thus the request of the display equipment's brightness is higher. While polarization 3D display technology has better imaging effect comparing to anaglyphic 3D display technology, its cost is not too much as well. The technology is used widely in most cinemas at present.

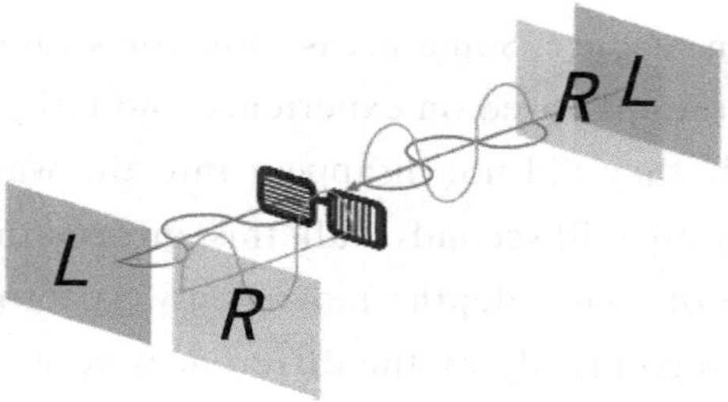

Figure 9-17 The course of redrawing stereo images with polarization 3D glasses

Polarization 3D uses the principle of vibration direction to disintegrate the original image. Firstly it divides images into vertical polarized light and level polarized light two groups images, then the left and right 3D glasses adopt polarized lenses with different polarized direction respectively. As Figure 9-17 shows the course redrawing stereo images with polarization 3D glasses, people's eyes can receive two groups of pictures, and then combine stereo images together through brains.

We can see that when stacking the same direction lens, euphotic rate decreases obviously; while stacking the vertical direction lens, it is far from to transmit light in polarization 3D glasses.

(3) Active shutter 3D

Active shutter 3D display technology is widely applied in television and the projector. It has many resources relatively, and its image effect is outstanding. Though the matching 3D glasses cost high expenses, it is highly praised by a lot of manufactures.

Active shutter 3D achieves 3D effect mainly through improving the refresh rate of pictures. It puts the image in two by frame, forms two pictures corresponding to the left and right eyes, and interlaces to display continuously. At the same time the infrared signal transmitters will control the left and right lens switch of active shutter 3D glasses synchronously, making the left and right eyes see the corresponding picture at the right moment. This technology can keep the original picture resolution, make users enjoy the real full high definition (HD) 3D effect very easily, and won't reduce screen brightness.

9.4.3 Free Viewing

With practice, most readers can view stereo pairs without the aid of blocking devices using a technique called free viewing. There are two types of free viewing, distinguished by how the left and right eye images are arranged. In parallel or uncrossed viewing, the left eye image is to the left of the right eye image. In transverse or cross viewing, they are reversed and crossing the eyes to form an image in the center is required. Some people can do both types of viewing, some only one, some neither. In Figure 9-18, the eye views have been arranged in left/right/left order. To parallel view, look at the left two images. To cross view, look at the right two images.

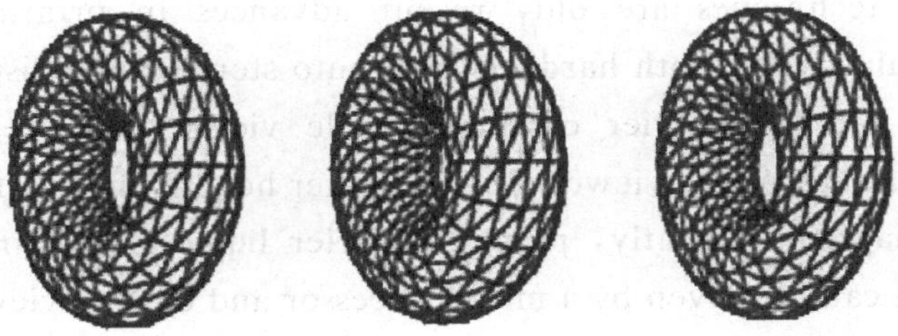

Figure 9-18 Free viewing examples

(1) Parallax barrier displays

A parallax barrier consists of a series of fine vertical slits in an otherwise opaque medium. The barrier is positioned close to an image that has been recorded in vertical slits and back lit. If the vertical slits in the image have been sampled with the correct frequency relative to the slits in the parallax barrier, and the viewer is at the required distance from the barrier, then the barrier will occlude the appropriate image slits to the right and left eyes respectively and the viewer will perceive an autostereoscopic image. The images can be made panoramic to some extent by recording multiple views of a scene. As the viewer changes position, different views of the scene will be directed by the barrier to the visual system. The number of views is limited by the optics and, hence, moving horizontally beyond a certain point will produce "image flipping" or a cycling of the different views of the scene.

High resolution laser printing has made it possible to produce very high quality images: the barrier is printed on one side of a transparent medium and the image on the other. This technique was pioneered in the early 1990s to produce hard copy displays.

(2) Lenticular lens

Figure 9-19 shows the principle of lenticular lens display. A lenticular sheet consists of a series of semi-cylindrical vertical lenses called "lenticles," typically made of plastic. The

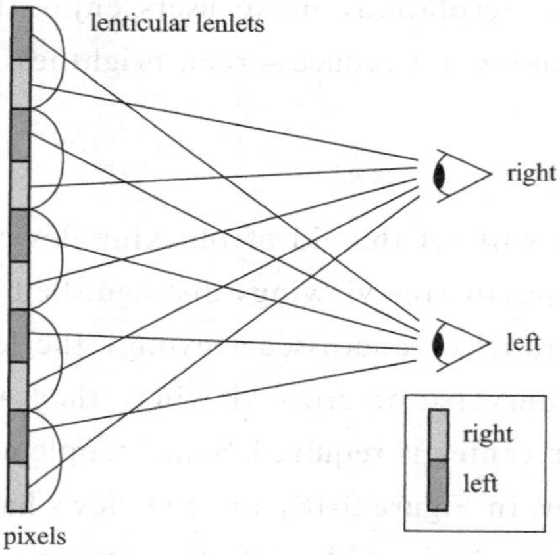

Figure 9-19 The principle of lenticular lens display

sheet is designed so the parallel light entering the front of the sheet will be focused onto strips on the flat rear surface. By recording an image in strips consistent with the optics of the lenticles as in the case of the parallax barrier display, an autostereoscopic panoramic image can be produced. Because the displays depend on refraction vs. occlusion, the brightness of a lenticular sheet display is usually superior to the parallax barrier and requires no back-lighting. Such displays have been mass produced for many years for such hardcopy media as postcards.

In these two techniques the image is recorded in strips behind the parallax barrier or the lenticular sheet. Although the techniques are old, recent advances in printing and optics have increased their popularity for both hardcopy and auto stereoscopic display devices. In both the lenticular and parallax barrier cases, multiple views of a scene can be included providing motion parallax as the viewer moves his/her head from side to side creating what is called a panoramagram. Recently, parallax barrier liquid-crystal imaging devices have been developed that can be driven by a microprocessor and used to view stereo pairs in real time without glasses.

(3) Holographic stereogram

Most readers are familiar with holographic displays, which reconstruct solid images. Normally a holographic image of a three dimensional scene will have the "look around" property. A popular combination of holography and stereo pair technology, called a holographic stereogram, involves recording a set of 2D images, often perspective views of a scene, on a piece of holographic film. The film can be bent to form a cylinder so the user can walk around the cylinder to view the scene from any aspect. At any point the left eye will see one view of the scene and the right eye another or the user is viewing a stereo pair.

Conventional display holography has long been hampered by many constraints such as limitations with regard to color, view angle, subject matter limitations, and final image size. Even with the proliferation of holographic stereogram techniques in the 1980s, the majority of the constraints remained. Zebra Imaging, Inc. expanded on the developments in one-step holographic stereogram printing techniques and has developed the technology to print digital full-color reflection holographic stereograms with a very wide view angle (up to 110°), unlimited in size, and with full parallax.

Zebra Imaging's holographic stereogram technique is based on creating an array of small (1 mm or 2 mm) square elemental holographic elements. Much like the pixels of two-dimensional digital images, hogel (short for holographic pixel) arrays can be used to form complete images of any size and resolution. Each hogel is a reflection holographic recording on pan-chromatic photopolymer film. The image recorded in each hogel is of a two-dimensional digital image on a spatial light modulator (SLM) illuminated with laser light in the three primary colors: red, green, and blue.

New Words and Expressions

a stack of 一堆，许多
cascade 串联，级联
CAT scan 计算机断层扫描技术
cellular 细胞的
census 人口调查，统计数字
chamber 腔，室
clinic 诊所，临床
condense 冷凝
deflect 偏转，偏移
debate 争论，辩论
deposition 沉积
extinction coefficient 消光系数
fabricate 建造
filter wheel 滤光片转盘
lifespan 寿命
malignant 恶性的
manifestation 显示，表现
penetrating 渗透，浸入
shrinking 萎缩
shutter 快门
superficial 表面的，肤浅的
synthesize 合成
therapy 疗法
thin film 薄膜
thinning agent 稀释剂

Translation

1. Filters are composed of a stack of such layers, alternating between high and low refractive indices, with typically 20 ～ 40 layers, although in some cases they may have one hundred layers or more.

翻译： 滤光片由一系列这样的层组成，在高折射率和低折射率之间交替变化，尽管在某些情况下它们可能有 100 层或更多层，但是通常只有 20～40 层。

2. Antireflection coatings operate on the principle of interference of the light reflected from the front and back surfaces of the films.

翻译： 减反射膜是根据薄膜前后表面反射光的干涉原理工作的。

3. The bandpass filter transmits a relatively narrow range of wavelengths and effectively rejects those on both sides.

翻译： 带通滤波器传输相对较窄的波长范围，并有效地抑制两侧的波长。

4. Since the effect is inversely proportional to the film index, at large angles the layers are no longer matched at the same wavelength and the spectral shape degrades.

翻译： 由于这种效应与薄膜折射率成反比，在大角度下，各层在相同波长下不再匹配，光谱图形也会递降。

5. To improve thickness uniformity, the holder is often a dome which is rotated about the vertical axis of the chamber.

翻译： 为了提高厚度均匀性，保持架通常是一个圆顶，它绕着光谱室的垂直轴旋转。

6. Since the forces which hold the film onto the substrate are all short range, even a few molecular layers of contaminant will result in a film with reduced adhesion.

翻译： 由于将薄膜固定在基底上的力都是短程的，即使是几层分子污染物也会导致薄膜的附着力降低。

7. Tighter control may often be achieved with planetary tooling where smaller plates are rotated about their axes which are also moved about the chamber axis in a planetary motion.

翻译：更严格的控制通常可以通过行星工具来实现，其中较小的板绕着它们的轴旋转，它们也在行星运动中绕着腔室轴移动。

8. As implied earlier, the optical thickness (index times physical thickness) must be controlled accurately for each layer.

翻译：如前所述，必须精确控制每层的光学厚度（指数乘以物理厚度）。

9. There have been many schemes devised for controlling thickness, but the optical monitor and the quartz crystal microbalance are the two methods which are most widely used today.

翻译：控制厚度的方法有很多种，但光学监测仪和石英晶体微天平是目前应用最广泛的两种方法。

10. Photolithography, or optical lithography is a process used in microfabrication to selectively remove parts of a thin film or the bulk of a substrate.

翻译：光刻或光学光刻是一种在微加工中用来选择性地去除薄膜的一部分或一块基板的过程。

11. Photonics is an all-encompassing light-based optical technology that is being hailed as the dominant technology for this new millennium.

翻译：光子学是一种全方位的基于光的光学技术，被誉为新千年的主导技术。

12. As an increasingly aging world population presents unique health problems, biophotonics offers great hope for the early detection of diseases and for new modalities of light-guided and light-activated therapies.

翻译：随着日益老龄化的世界人口呈现出独特的健康问题，生物光子学为疾病的早期检测以及光引导和光激活疗法的新模式带来了巨大的希望。

References

[1] 张彬. 光电信息科学与工程专业英语教程 [M]. 北京：电子工业出版社，2012.

[2] 张爱红. 电子科学与技术专业英语 光电信息技术分册 修订版 [M]. 哈尔滨：哈尔滨工业大学出版社，2017.

[3] 徐朝鹏，王朝晖，焦斌亮. 光电子技术专业英语 [M]. 北京：北京邮电大学出版社，2010.

[4] 钟似璇. 英语科技论文写作与发表 [M]. 天津：天津大学出版社，2004.

[5] 辜嘉铭. 英语科技论文写作精要 [M]. 武汉：武汉大学出版社，2006.

[6] 李霞，王娟. 电子与通信专业英语 [M]. 北京：电子工业出版社，2010.

[7] 金志权，张幸儿. 计算机专业英语教程 [M]. 北京：电子工业出版社，2010.

[8] 毛样武. 材料物理科技英语 [M]. 北京：知识产权出版社，2015.

[9] 王高，姚密红. 电子科学技术专业英语与科技论文写作 [M]. 北京：兵器工业出版社，2005.

[10] 闫小兵，师建英. 电子科学与技术专业英语 [M]. 北京：科学出版社，2019.

[11] 谢瑶. 信息技术英语读本 [M]. 昆明：云南大学出版社，2019.

[12] 李白萍. 电子信息类专业英语 [M]. 西安：西安电子科技大学出版社，2003.

[13] 王琳，夏怡. 电子与通信专业英语 [M]. 北京：北京理工大学出版社，2007.

[14] 姜宇. 通信电子信息类专业英语 [M]. 哈尔滨：东北林业大学出版社，2006.

[15] 王凌波，王九程. 光电技术应用专业英语 [M]. 武汉：华中科技大学出版社，2018.

[16] 刘建宇，魏巍. 光电世界 [M]. 北京：国防工业出版社，2004.

[17] 周彦平. 专业英语阅读丛书 电子科学与技术专业英语 [M]. 哈尔滨：哈尔滨工业大学出版社，2003.

[18] 苏雪主. 计算机专业英语 [M]. 武汉：华中科技大学出版社，2007.

[19] 马丽华，李瑞欣，王豆豆. 通信工程专业英语 [M]. 北京：北京邮电大学出版社，2009.

[20] 韩俊岗，袁立行，王忠民. 信息科学类专业英语 [M]. 西安：西安电子科技大学出版社，2011.

[21] 屈晶，卜玉坤. 大学专业英语 电气与电子英语 2 [M]. 北京：外语教学与研究出版社，2008.

[22] 王群. 电气信息类专业英语 [M]. 长沙：湖南大学出版社，2010.

[23] 常义林，任志纯. 通信工程专业英语 [M]. 西安：西安电子科技大学出版社，2001.

[24] 翟俊祥，杨向明. 信息与控制专业英语 [M]. 西安：西安交通大学出版社，2000.

[25] Young M. Optics and Lasers: Including Fibers and Optical Waveguides. 5th ed [M]. Berlin: Springer-Verlag Berlin Heidelberg, 2000.

[26] Eugene H. Optics. 4th ed [M]. New York: Adelphi University, 2001.

[27] Govind A. Nonlinear Fiber Optics. 3rd ed [M]. Pittsburgh: Academic Press, 2001.

[28] Amnon Y. Optical Electronics in Modern Communications. 5th ed [M]. Beijing: Publishing House of Electronics Industry, 2004.

[29] Max B, Emil W. Principles of Optics. 7th ed [M]. New York: Cambridge University Press, 1980.

[30] Hariharan P. Optical Holography: Principles, Techniques and Applications. 2nd ed [M]. New York: Press Syndicate of University of Cambridge, 1996.